Systems with Hysteresis

Systems with Hysteresis

Analysis, Identification and Control using the Bouc–Wen Model

Fayçal Ikhouane
Department of Applied Mathematics III
School of Technical Industrial Engineering
Technical University of Catalunya
Barcelona, Spain

José Rodellar
Department of Applied Mathematics III
School of Civil Engineering
Technical University of Catalunya
Barcelona, Spain

John Wiley & Sons, Ltd

Telephone (+44) 1243 779777

Email (for orders and customer service enquiries): cs-books@wiley.co.uk
Visit our Home Page on www.wiley.com

Other Wiley Editorial Offices

John Wiley & Sons Inc., 111 River Street, Hoboken, NJ 07030, USA

Jossey-Bass, 989 Market Street, San Francisco, CA 94103-1741, USA

Wiley-VCH Verlag GmbH, Boschstr. 12, D-69469 Weinheim, Germany

John Wiley & Sons Australia Ltd, 42 McDougall Street, Milton, Queensland 4064, Australia

John Wiley & Sons (Asia) Pte Ltd, 2 Clementi Loop #02-01, Jin Xing Distripark, Singapore 129809

John Wiley & Sons Canada Ltd, 6045 Freemont Blvd, Mississauga, ONT, L5R 4J3

Wiley also publishes its books in a variety of electronic formats. Some content that appears in print may not be available in electronic books.

Anniversary Logo Design: Richard J. Pacifico

Library of Congress Cataloging in Publication Data

Ikhouane, Fayçal.
Systems with hysteresis : analysis, identification and control using the Bouc-Wen model / Fayçal Ikhouane, José Rodellar.
p. cm.
Includes bibliographical references and index.
ISBN 978-0-470-03236-7 (cloth)
1. Hysteresis—Mathematical models. I. Rodellar, José. II. Title.
QC754.2.H9I34 2007
621—dc22

2007019894

British Library Cataloguing in Publication Data

A catalogue record for this book is available from the British Library

ISBN-13 978-0-470-03236-7

Typeset in 11/13pt Sabon by Integra Software Services Pvt. Ltd, Pondicherry, India
Printed and bound in Great Britain by TJ International, Padstow, Cornwall
This book is printed on acid-free paper responsibly manufactured from sustainable forestry in which at least two trees are planted for each one used for paper production.

To my mother and brothers, Imad and Hicham F. Ikhouane

To Anna, Laura and Silvia J. Rodellar

Contents

Preface

This book deals with the analysis, the identification and the control of a special class of systems with hysteresis. This nonlinear behaviour is encountered in a wide variety of processes in which the input–output dynamic relations between variables involve memory effects. Examples are found in biology, optics, electronics, ferroelectricity, magnetism, mechanics and structures, among other areas. In mechanical and structural systems, hysteresis appears as a natural mechanism of materials to supply restoring forces against movements and dissipate energy. In these systems, hysteresis refers to the memory nature of inelastic behaviour where the restoring force depends not only on the instantaneous deformation but also on the history of the deformation.

The detailed modelling of these systems using the laws of physics is an arduous task, and the obtained models are often too complex to be used in practical applications involving characterization of systems, identification or control. For this reason, alternative models of these complex systems have been proposed. These models do not come, in general, from the detailed analysis of the physical behaviour of the systems with hysteresis. Instead, they combine some physical understanding of the hysteretic system along with some kind of black-box modelling. For this reason, some authors have called these models 'semi-physical'.

Within this context, a hysteretic semi-physical model was proposed initially by Bouc early in 1971 and subsequently generalized by Wen in 1976. Since then, it is known as the Bouc–Wen model and has been extensively used in the current literature to describe mathematical components and devices with hysteretic behaviours, particularly within the areas of civil and mechanical engineering. The model essentially consists of a first-order nonlinear differential equation that relates the input displacement to the output restoring force in a hysteretic way. By choosing a set of parameters appropriately, it is possible to accommodate the response of the model to the real hysteresis loops. This is why the main efforts reported in the literature have been devoted to the tuning of the parameters for specific applications.

This book is the result of a research effort that was initiated by the first author (Prof. Fayçal Ikhouane) in 2002 when he joined the research group on Control, Dynamics and Applications (CoDAlab) in the Department of Applied Mathematics III at the Technical University of Catalonia in Barcelona (Spain). During the last five years, the authors have explored various issues related to this model as an analysis of some physical properties of the model, and the parameteric identification and control of systems that include the Bouc–Wen model.

The book has been written to compile the results of this research effort in a comprehensive and self-contained organized body. Part of these results have been published in scientific journals and presented in international conferences within the last three years as well as in lectures and seminars for graduate students. The contents cover four topics:

1. Analysis of the compatibility of the model with some laws of physics.
2. Relationship between the model parameters and the hysteresis loop.
3. Identification of the model parameters.
4. Control of systems that include a Bouc–Wen hysteresis.

Although mathematical rigour has been the main pursued feature, the authors have also tried to make the book attractive for, say, end users of the model. Thus, the mathematical developments are completed with practical remarks and illustrated with examples. Their final goal is that the analytical studies and results give a solid framework for a systematic and well-supported practical use of the Bouc–Wen model. It is their hope that this has been achieved and that the book might be of interest to researchers, engineers, professors and students involved in the design and development of smart structures and materials, vibration control, mechatronics, smart actuators and related issues in engineering areas such as civil, mechanical, automotive, aerospace and aeronautics.

The research work leading to this book has been mainly sponsored by the Ministry of Education and Science of Spain through research project grants. Particularly significant is the grant 'Ramón y Cajal' awarded to Prof. F. Ikhouane for the last five years. Additional support from the Research Agency of the Government of Catalonia is appreciated. This work has also benefited from the participation of the group CoDAlab in the program CONVIB (Innovative Control Technologies for Vibration Sensitive Civil Engineering Structures) sponsored by the European Science Foundation (ESF) during the period 2001–2005.

The School of Industrial Technical Engineering of Barcelona (EUETIB) and the School of Civil Engineering of Barcelona (ETSECCPB) have provided a pleasant environment for our work. There is also appreciation for the support of colleagues and graduate students at the Technical University of Catalonia.

We are grateful to Prof. Shirley Dyke, Prof. Víctor Mañosa and Prof. Jorge Hurtado for their coauthoring of some of the papers that have been used in this book, and also to anonymous reviewers for their valuable comments.

Finally, we owe a special gratitude for the permanent support of our respective families during the research work and the writing of this book.

Fayçal Ikhouane and José Rodellar

List of Figures

List of Tables

1

Introduction

1.1 OBJECTIVE AND CONTENTS OF THE BOOK

Hysteresis is a nonlinear phenomenon exhibited by systems stemming from various science and engineering areas: under a low-frequency periodic excitation, the relationship between the system's input and output is not the same for loading and unloading. More precisely, consider a single-input single-output (SISO) system excited by a periodic signal that has a loading–unloading shape. Then, hysteretic systems often present a periodic response that has the same frequency of the input. When this frequency goes to zero, the quasi-static response of the system has an output versus input plot that is a cycle (not a line as would be the case for linear systems).

A fundamental theory allowing a general mathematical framework for modelling hysteresis has not been developed up to now. For specific problems, models describing hysteretic systems can be derived from an understanding of physical laws. Usually this is an arduous task and the resulting models are too complex to be used in practical applications. In general, engineering practice seeks for alternative more simple models which, although not giving the 'best' description of the physical behaviour of the system, do keep relevant input–output features and are useful for characterization, design and control purposes. These models are referred to as phenomenological or semi-physical models.

Systems with Hysteresis: Analysis, Identification and Control using the Bouc–Wen Model
F. Ikhouane and J. Rodellar

In this context, several mathematical models have been proposed to describe the behaviour of hysteretic processes [1]. The Duhem model [2] uses the property that a hysteretic system's output changes its character when the input changes direction; the Ishlinskii hysteresis operator has been proposed as a model for plasticity–elasticity [3] and the Preisach model has been used for modelling electromagnetic hysteresis [4]. A survey of mathematical models for hysteresis may be found in [5]. In the areas of smart structures and civil engineering, another model has been used extensively to describe the hysteresis phenomenon: the so-called Bouc–Wen model [6,7]. It consists of a first-order nonlinear differential equation that relates the input displacement to the output restoring force in a rate-independent hysteretic way. The parameters that appear in the differential equation can be tuned to match the hysteresis loop of the system under study.

The current literature devoted to the Bouc–Wen model is extensive and focuses mainly on:

1. Tuning the model parameters to obtain a reasonable matching of the physical hysteretic system under consideration.
2. Use of the obtained tuned model for simulation and control purposes.

It is known that most works on this model have been practically oriented. In general, rigorous mathematical justifications of the techniques associated with the use of the model have been missing. To give an example, while many papers have been devoted to tuning the Bouc–Wen model parameters (that is the identification problem), rigorous proofs on the convergence of the identified model parameters to their true counterparts are still lacking. Most works rely mainly on numerical simulations to show this convergence.

The objective of this book is to contribute to fill this gap by providing the reader with a rigorous treatment of this model. This book is based on original works by the authors that have been published in scientific journals within the last three years. It includes a mathematical treatment of the subject along with several numerical simulation examples. The book covers basically four topics:

1. Analysis of the compatibility of the model with some laws of physics.

2. Relationship between the model parameters and the hysteresis loop.
3. Identification of the model parameters.
4. Control of systems that include a hysteretic part described by the Bouc–Wen model.

The first topic is about checking whether the semi-physical Bouc–Wen model is consistent with some general laws of physics. In particular, the conditions are given under which the model is input–output stable and passive. These conditions translate into inequalities that have to be satisfied by the Bouc–Wen model parameters in order to comply with the stability and the passivity properties. Also cited is a parallel work by other authors that checks the thermodynamical admissibility of the Bouc–Wen model [8]. The techniques used in this part of the book include Lyapunov techniques for checking the stability of the model and passivity methods for the analysis of energy dissipation. The result of this analysis is a set of inequalities to be held by the Bouc–Wen model parameters. These inequalities will prove to be fundamental in deriving a new form of the Bouc–Wen model that can be called the normalized one. This new form will be used extensively in the rest of the book. This first topic is the subject of Chapter 2.

The second topic is the subject of Chapters 3 and 4. Chapter 3 is devoted to the analytical description of the hysteresis loop. Indeed, it is well known that, under loading and unloading, physical systems with hysteresis do not follow the same path, which results in a hysteresis loop. Due to the nonlinearity of the Bouc–Wen model, the hysteresis loop has never been described analytically in an explicit way. This lack of knowledge has impeded analytical studies on the relationship between the model parameters and the shape and size of the hysteresis loop. Chapter 3 presents a novel result of the authors where, using a simple but rigorous mathematical framework, the hysteresis loop is described analytically using some explicit functions that can be computed numerically in an easy way. This analytical description is illustrated and commented upon by means of a numerical simulation example.

Chapter 4 uses the analytical description of Chapter 3 to study the behaviour of the hysteresis loop when the Bouc–Wen model parameters change. This chapter is basically divided into two parts. The first part is focused on the variation of a given point of the hysteresis loop along the axes of abscissas and ordinates when the parameters of the normalized Bouc–Wen model vary. The results of this part are summarized in tables to facilitate their use. In the second part of Chapter 4, the hysteresis loop of the Bouc–Wen model is divided into four regions: the linear region, the plastic region and two regions of transition. The points that define each region are defined rigorously, which allows an analysis of the behaviour of the different regions with respect to the normalized Bouc–Wen model parameters. These regions are illustrated by means of several figures.

The third topic is the subject of Chapter 5. Identification of the parameters of the Bouc–Wen model is a crucial issue and a technical challenge for its practical use. This issue has been treated in the literature using numerical simulations, and, to the best of the authors' knowledge, no currently available method ensures analytically that the identified parameters converge to their true counterparts. In this chapter, a new identification technique is presented that uses the results of Chapter 3 to identify in an exact way the parameters of the normalized Bouc–Wen model. The technique consists of imposing two specific input displacement functions that are wave-periodic; this means that the displacements have a loading–unloading shape, and are periodic in time. Then the two obtained limit cycles are used to identify the Bouc–Wen model parameters. Chapter 5 is divided into two parts:

1. The first part presents the identification methodology and analyses its robustness with respect to external disturbances.
2. The second part of the chapter consists in applying this methodology to a magnetorheological (MR) damper, which is described by a model that includes a Bouc–Wen hysteresis. The values of the parameters of the model are taken from the literature and are unknown to the identification algorithm. Numerical simulations are carried out to illustrate the applicability of the identification method.

The fourth topic is the subject of Chapter 6. It consists of the control of a mechanical/structural system containing a hysteresis described by the Bouc–Wen model, and represents a base-isolated structure. The system parameters are not known exactly but they lie in known intervals. The control objective is to regulate the system around zero while maintaining the boundedness of the closed-loop signals. The control law is a simple proportional-integral-derivative

(PID) whose parameters are to be tuned in a specific way to guarantee the boundedness of all the closed-loop signals. Furthermore, the controller ensures the asymptotic convergence to zero of the mass displacement and velocity. The interest of this chapter is to show that a linear controller may ensure the control objective in the presence of a Bouc–Wen hysteresis.

1.2 THE BOUC–WEN MODEL: ORIGIN AND LITERATURE REVIEW

The starting point of the so-called Bouc–Wen model is the early paper by Bouc [6], where a functional that describes the hysteresis phenomenon was proposed. Consider Figure 1.1, where $\mathcal{F}$ is a force and x a displacement. Four values of $\mathcal{F}$ correspond to the single point $x = x_0$, which means that $\mathcal{F}$ is not a function. If it is considered that x is a function of time, then the value of the force at the instant time t will depend not only on the value of the displacement x at the time t, but also on the past values of x. The following simplifying assumption is made in Reference [6].

Assumption 1. *The graph of Figure 1.1 remains the same for all increasing functions $x(\cdot)$ between 0 and x_1, for all decreasing functions $x(\cdot)$ between the values x_1 and x_2, etc.*

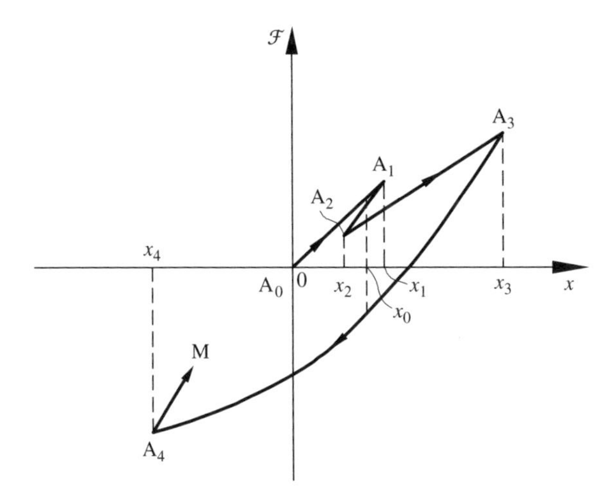

Figure 1.1 Graph force versus displacement for a hysteresis functional.

Assumption 1 is what, in the current literature, is called the *rate-independent property* [1]. To define the form of the functional $\mathcal{F}$, Reference [6] elaborates on previous works to propose the following form:

$$\frac{\mathrm{d}\mathcal{F}}{\mathrm{d}t} = g\left(x, \mathcal{F}, \operatorname{sign}\left(\frac{\mathrm{d}x}{\mathrm{d}t}\right)\right)\frac{\mathrm{d}x}{\mathrm{d}t} \tag{1.1}$$

Consider the equation

$$\frac{\mathrm{d}^2x}{\mathrm{d}t^2} + \mathcal{F}(t) = p(t) \tag{1.2}$$

for some given input $p(t)$ and initial conditions

$$\frac{\mathrm{d}x}{\mathrm{d}t}(t_0), \qquad x(t_0) \qquad \text{and} \qquad \mathcal{F}(t_0)$$

at the initial time instant t_0. Equations (1.1) and (1.2) describe completely a hysteretic oscillator.

Paper [6] notes that it is difficult to give explicitly the solution of Equation (1.1) due to the nonlinearity of the function g. For this reason, the author proposes the use of a variant of the Stieltjes integral to define the functional $\mathcal{F}$:

$$\mathcal{F}(t) = \mu^2 x(t) + \int_\beta^t F(V_s^t)\,\mathrm{d}x(s) \tag{1.3}$$

where $\beta \in [-\infty, +\infty)$ is the time instant after which the displacement and force are defined. The term V_s^t is the total variation of x in the time interval $[s, t]$. The function F is chosen in such a way that it satisfies some mathematical properties compatible with the hysteresis property. The following is an example of this choice given in Reference [6] so that these mathematical properties are satisfied:

$$F(u) = \sum_{i=1}^{N} A_i \mathrm{e}^{-\alpha_i u} \qquad \text{with } \alpha_i > 0 \tag{1.4}$$

Equations (1.2) to (1.4) can then be written in the form

$$\frac{\mathrm{d}^2x}{\mathrm{d}t^2} + \mu^2 x + \sum_{i=1}^{N} Z_i = p(t) \tag{1.5}$$

$$\frac{\mathrm{d}Z_i}{\mathrm{d}t} + \alpha_i \left| \frac{\mathrm{d}x}{\mathrm{d}t} \right| Z_i - A_i \frac{\mathrm{d}x}{\mathrm{d}t} = 0, \qquad i = 1, \ldots, N \tag{1.6}$$

Equations (1.5) and (1.6) are what is now known as the Bouc model. The derivation of these equations is detailed in Reference [6]. The objective here is not to enter in these details, but only to give a short idea of the origin of the model. Equation (1.6) has been extended in Reference [7] to describe restoring forces with hysteresis in the following form:

$$\dot{z} = -\alpha|\dot{x}|z^n - \beta\dot{x}|z^n| + A\dot{x} \qquad \text{for } n \text{ odd} \tag{1.7}$$

$$\dot{z} = -\alpha|\dot{x}|z^{n-1}|z| - \beta\dot{x}z^n + A\dot{x} \qquad \text{for } n \text{ even} \tag{1.8}$$

Equations (1.7) and (1.8) constitute the earliest version of what is now called the Bouc–Wen model. The shape of the hysteresis loop is given in Reference [7] for different values of the model parameters. Some subsequent works have proposed different modifications of the model to take into account some physical properties observed experimentally in some hysteretic systems. In Reference [9], the authors consider the modelling of degradation in civil engineering structures. A multidegree of freedom shear beam structure is modelled in the form

$$m_i \left(\sum_{j=1}^{i} \ddot{u}_j + \ddot{\xi}_B \right) + q_i - q_{i+1} = 0 \qquad \text{for } i = 1, \ldots, n \tag{1.9}$$

in which m_i is the mass of the ith floor, $\ddot{\xi}_B$ is the ground acceleration and q_i is the ith restoring force, including viscous damping. The quantities u_i are the relative displacement of the ith and the $(i+1)$th stories, and q_i is given as

$$q_i = c_i\dot{u}_i + \alpha_i k_i u_i + (1-\alpha_i)k_i z_i \qquad \text{for } i = 1, \ldots, n \tag{1.10}$$

in which c_i is the viscous damping, k_i controls the initial tangent stiffness, α_i controls the ratio of post-yield to pre-yield stiffness and z_i is the ith hysteresis which obeys the equation

$$\dot{z}_i = \frac{A_i\dot{u}_i - \nu_i\left(\beta_i|\dot{u}_i||z_i|^{n_i-1}z_i + \gamma_i\dot{u}_i|z_i|^{n_i}\right)}{\eta_i} \qquad \text{for } i = 1, \ldots, n \tag{1.11}$$

where A_i, ν_i, β_i, γ_i, η_i and n_i are parameters that control the hysteresis shape and the degradation of the system. System degradation is introduced into the model for z_i by allowing the parameters of the model to vary as a function of the response duration and severity. Pinching has been considered in Reference [10] by modifying the Bouc–Wen model in the form

$$\dot{z} = h(z)\,\frac{\dot{u} - \nu\left(\beta|\dot{u}||z|^{n-1}z + \gamma\dot{u}|z|^{n}\right)}{\eta} \tag{1.12}$$

where $h(z)$ is the function that describes pinching. A discussion on how to choose this function for wood systems is given in Reference [11]. Other modifications of the Bouc–Wen model include ones to describe a soft soil [12], an asymmetric response as observed in shape memory alloys [13], the response of steel buildings under earthquakes [14], the drift observed under a zero-mean, broad-band, stationary-random load [15] and the behaviour of low yield strength steel [16]. In a parallel research line, extensions of the Bouc or Bouc–Wen models to the multivariate case have been done in References [17] and [18].

Figure 1.2 illustrates that the literature on the Bouc–Wen model has increased rapidly during the last few years. It quantifies the number of papers published in journal papers, most of which are quoted in the references given at the end of the book.

One of the main issues in the literature devoted to the Bouc–Wen model is parameter identification. Several techniques have been

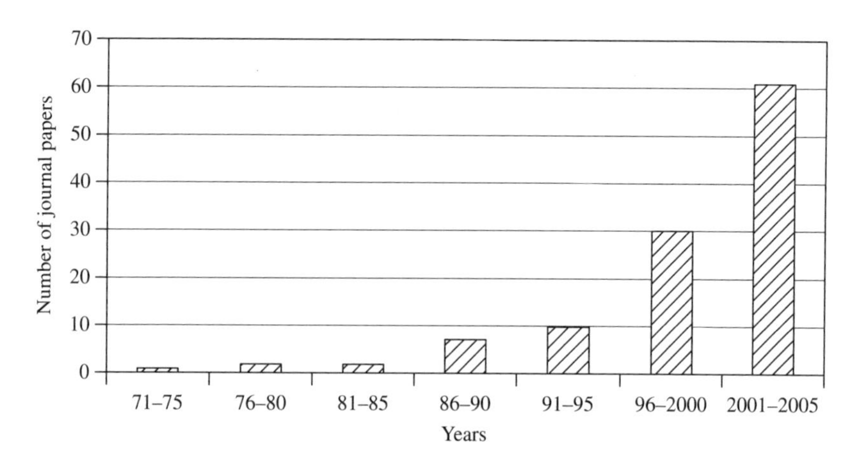

Figure 1.2 Evolution of the Bouc–Wen model literature.

used to deal with this problem. In Reference [19], a nonrecursive least error minimization algorithm is used. A recursive least-squares algorithm has been used in Reference [20], along with the Newton method and the extended Kalman filtering technique. More recent works that use some version of the least-squares algorithm include References [21] to [25]. For example, Reference [21] considers a second-order single-degree-of-freedom system which is a mass subject to a nonlinear restoring force and an external excitation. The restoring force is represented as a Bouc–Wen hysteresis whose input is the velocity of the mass. When the mass is exactly known, the restoring force can be calculated knowing the instantaneous external excitation and the acceleration of the mass. In this case, all the Bouc–Wen model parameters appear linearly except the exponent of the differential equation. This nonlinearity is coped with by assuming knowledge of an upper bound on the exponent and writing the Bouc–Wen differential equation as a sum of terms whose number is the upper bound. Then, a first-order filter is used to write the nonlinear system in a way that allows the use of the least-squares algorithm to identify the system parameters. The case of unknown mass is treated similarly by using an on-line estimation of the restoring force.

Genetic-type algorithms for the determination of the Bouc–Wen model parameters have been used in References [26] to [29]. For example, Reference [27] uses a differential evolution algorithm whose main difference with conventional genetic algorithms is in the way the mechanisms of mutations and crossover are performed using real floating point numbers instead of long strings of zeros and ones. This algorithm starts with an initial pool of 15 three-dimensional vectors drawn from uniform probability distributions. The differential evolution mutates a randomly selected number of the featured generation with vector differentials. Each differential is the difference between two randomly selected vectors, scaled with a parameter. This process generates a new mutated vector. Natural selection is implemented via a comparison process between the cost of the trial vector and the cost of the target vector. The differential evolution algorithm generates a new set of 15 three-dimensional vectors, which is a new generation with improved characteristics.

Methods that use the frequency domain have been utilized in References [30] to [33]. For example, Reference [30] considers a second-order system coupled with a Bouc hysteresis. The nonlinear system is excited with a periodic input and the Bouc model parameters are determined by using a first harmonic approximation. A higher number of harmonics is considered in Reference [31].

Neural networks have been used in Reference [34]. In this work, an inverse model for a magnetorheological damper has been developed using a multilayer perception network and system identification-based ARX model.

Bayesian parameter estimation is used in References [35] to [38]. For example, Reference [35] uses a modified version of the extended Kalman filter and the particle filter to determine the parameters of a second-order Bouc–Wen hysteresis.

A nonparametric identification method has been proposed in Reference [39]. The nonlinear hysteresis part of the system is written as a linear combination of polynomial functions with unknown coefficients. These coefficients are determined using a least-squares algorithm.

Other proposed identification techniques are included in References [40] to [48].

Control of mechanical systems and structures with Bouc–Wen hysteretic behaviour has also spurred much effort in the current literature. In this sense, it may be useful to distinguish between active and semi-active control. A control law is said to be active when the control signal directly feeds an actuator that applies the desired feedback control force. With an active control scheme, energy is injected into the closed-loop system. A control law is semi-active when the corresponding actuator does not pour energy into the closed loop. Instead, the control signal is generated by the controller to modify the characteristics of an adaptive passive-like actuator. Examples of semi-active actuators are the devices based on smart materials, in particular the magnetorheological dampers.

Now a brief overview of the recent control literature related to the Bouc–Wen model is given. Active control is described in References [49] to [58]. In Reference [49] fuzzy control is used for a structure modelled as a second-order single-degree-of-freedom structural system that includes a Bouc–Wen hysteresis. In Reference [51], an H_∞ controller is proposed to cope with the presence of uncertainties. In the other references nonlinear controllers based on Lyapunov techniques are used to ensure stability and some degree of performance in spite of the uncertainties.

Semi-active control is often used in relation to MR dampers. Reference [59] gives a state-of-the-art review of semi-active control systems for the seismic protection of structures. Recent references include [53] and [60] to [72]. For example, Reference [60] considers several semi-active control strategies using MR dampers for the control of a six-storey building. These control algorithms include a Lyapunov controller, decentralized bang-bang controller, modulated homogeneous friction algorithm and a clipped optimal controller. Each algorithm uses measurements of the absolute acceleration and device displacements for determining the control action to ensure that the algorithms would be implementable on a physical structure. The performance of the algorithms is compared through a numerical example, and the advantages of each algorithm are discussed.

The Bouc–Wen model has been extensively used for modelling hysteresis in structural and mechanical systems [44, 62, 73–95]. For example, Reference [88] considers an MR damper for which a dynamic model is to be developed. The damper force is written as the sum of several terms:

1. The damper friction due to seals and measurement bias.
2. The product of the equivalent mass which represents the MR fluid stiction phenomenon and inertial effect, and the acceleration of the piston.
3. The product of the piston velocity and the post-yield plastic damping coefficient.
4. The product of the piston position and the factor that accounts for the accumulator stiffness and the MR fluid compressibility.
5. A hysteretic term.

The hysteresis part of the model is assumed to follow a Bouc–Wen equation. Experiments are carried out to verify the validity of the model.

There are other works that have used the Bouc–Wen model [8, 96–123]. These works are difficult to classify into a single homogeneous group as their research subjects are diverse. However, they mostly deal with the analysis of some properties of systems that include a Bouc–Wen hysteresis. For example, Reference [107] analyses the influence of hysteresis dissipation on chaotic responses, Reference [113] studies the nonlinear response of a Bouc–Wen hysteretic oscillator under evolutionary excitation and Reference [110] addresses strategies for finding the design point in nonlinear finite element reliability analysis.

This book treats the univariate basic Bouc–Wen model, that is the one that has one input and one output, and describes only the hysteresis phenomenon regardless of other types of nonlinear behaviours (like pinching and others). This choice is motivated by the fact that most references treat only this basic Bouc–Wen model. The extension of the results of this book to the multivariate model, which may include other types of nonlinearities, is still an open problem and a possible subject for future research.

2

Physical Consistency of the Bouc–Wen Model

2.1 INTRODUCTION

In the current literature, the Bouc–Wen model is mostly used within the following black-box approach: given a set of experimental input–output data, how can the Bouc–Wen model parameters be adjusted so that the output of the model matches the experimental data? The use of system identification techniques is one practical way to perform this task. Once an identification method has been applied to tune the Bouc–Wen model parameters, the resulting model is considered as a 'good' approximation of the true hysteresis when the error between the experimental data and the output of the model is small enough. Then this model is used to study the behaviour of the true hysteresis under different excitations.

By doing this, it is important to consider the following remark. It may happen that a Bouc–Wen model presents a good matching with the experimental real data for a specific input, but does not necessarily keep significant physical properties that are inherent to the real data, independently of the exciting input. In this chapter attention is drawn to this issue, with particular focus on the following two properties, which are shared by most of the hysteretic mechanical and structural systems.

Property 1. Conceptualize a nonlinear hysteretic behaviour as a map $x(t) \mapsto \Phi_s(x)(t)$, where x represents the time history of an

Systems with Hysteresis: Analysis, Identification and Control using the Bouc–Wen Model
F. Ikhouane and J. Rodellar

input variable and $\Phi_s(x)$ describes the time history of the hysteretic output variable. For any bounded input x, the output of the true hysteresis $\Phi_s(x)$ is bounded. This bounded input–bounded output (BIBO) stability property stems from the fact that mechanical and structural systems are being dealt with, which are stable in the open loop.

Property 2. Consider that x is the displacement of a one-degree-of-freedom mechanical system connected to an element or device that supplies a hysteretic restoring force $\Phi_s(x)$ to the system. The hysteretic element or device contributes to dissipate the mechanical energy of the system as usually observed in practice. The Bouc–Wen model has to reproduce this energy dissipation property in order to represent adequately the physical behaviour of real systems.

Figure 2.1 shows an example of a typical hysteretic loop (x, Φ) obtained by the Bouc–Wen model for a specific set of parameters and for the signal $x(t) = \sin(t)$. However, Example 1 shows that other different bounded time histories x exist for which this Bouc–Wen model delivers unbounded responses $\Phi(x)$, which means that this model is not BIBO stable.

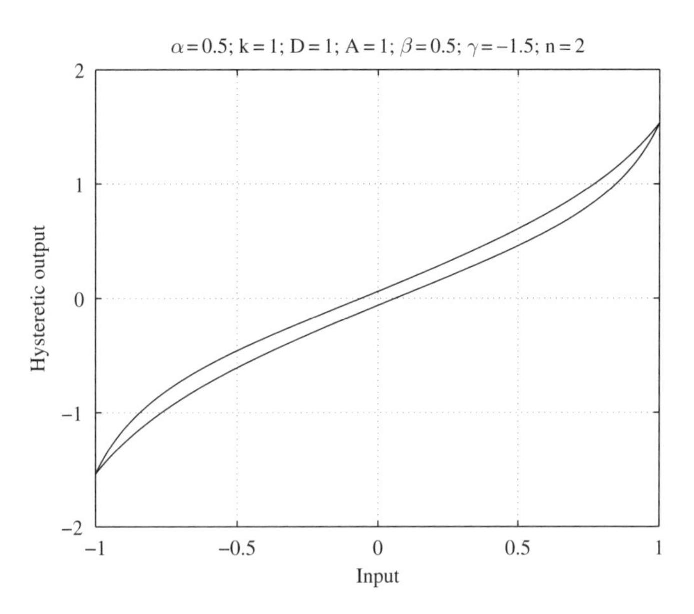

Figure 2.1 Example of a Bouc–Wen model that is unstable.

Example 1. *Consider the Bouc–Wen model of Equations (2.4) and (2.5) given by the following parameters*: $D=1,\ A=1,\ \beta=0.5,\ \gamma=-1.5$ *and* $n=2$. *Take* $z(0)=0$ *and define the bounded input signal* $x(t)=(\pi/2)\sin(t)$. *The corresponding derivative is* $\dot{x}(t)=(\pi/2)\cos(t)$, *which is also bounded. For* $0\leq t\leq\pi/2$, *then* $\dot{x}(t)\geq 0$. *This implies that, during the time interval* $[0,\pi/2]$, *the Bouc–Wen model (2.5) can take only one of the two forms*:

$$\dot{z}=\dot{x}\left(1+z^2\right) \qquad \text{for } z\geq 0 \tag{2.1}$$

$$\dot{z}=\dot{x}\left(1+2z^2\right) \qquad \text{for } z\leq 0 \tag{2.2}$$

In both cases (2.1) and (2.2), $\dot{z}\geq 0$ *for* $0\leq t\leq\pi/2$, *which implies that* $z(t)$ *is a nondecreasing function. Since* $z(0)=0$, *this means that* $z(t)\geq 0$, *so that* $\dot{z}$ *is given by (2.1). Integrating (2.1) gives*

$$\int\frac{\mathrm{d}z}{1+z^2}=\int\mathrm{d}x \tag{2.3}$$

which gives $\arctan(z)=x$, *since* $z(0)=0$ *and* $x(0)=0$. *This implies that* $z(t)=\tan[x(t)]$. *Observe that* $\lim_{t\to\pi/2}z(t)=+\infty$. *Thus the bounded input signal* $x(t)$ *has given rise to an unbounded hysteretic output. A similar construction can be done for any initial condition* $z(0)\neq 0$.

In a similar vein, the Bouc–Wen model illustrated in Figure 2.2 is BIBO stable. However, it can be shown that it does not dissipate the mechanical energy of the system as considered above in Property 2. These two examples highlight the fact that, while these models may give a good approximation of a true hysteresis loop for a *specific* input excitation used with parametric identification or tuning purposes, they may not be appropriate to represent the behaviour of a true hysteretic system under general input excitations.

This chapter presents an analytical study with the aim of giving the conditions on the Bouc–Wen model so that it holds the above BIBO stability and dissipation properties. The study uses mathematical tools related to system analysis, such as differential equations, stability theory and passivity. Some of these tools are summarized in the Appendix at the end of the book.

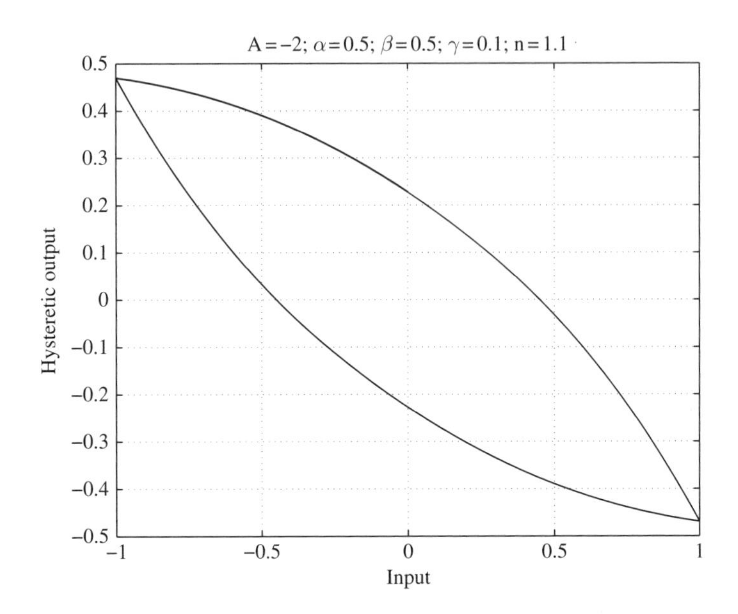

Figure 2.2 Example of a Bouc–Wen model that does not dissipate energy.

2.2 BIBO STABILITY OF THE BOUC–WEN MODEL

2.2.1 The Model

Consider a physical system with a hysteretic component that can be represented by a map $x(t) \mapsto \Phi_s(x)(t)$, which is referred to as the 'true' hysteresis. The so-called Bouc–Wen model represents the true hysteresis in the form

$$\Phi_{\mathrm{BW}}(x)(t) = \alpha k x(t) + (1-\alpha) D k z(t) \tag{2.4}$$

$$\dot{z} = D^{-1}\left(A\dot{x} - \beta|\dot{x}|\,|z|^{n-1}z - \gamma\dot{x}|z|^{n}\right) \tag{2.5}$$

where $\dot{z}$ denotes the time derivative, $n > 1$, $D > 0$, $k > 0$ and $0 < \alpha < 1$ (the limit cases $n = 1$, $\alpha = 0$, $\alpha = 1$ are treated in Section 2.5). It is also considered that $\beta + \gamma \neq 0$, the singular case $\beta + \gamma = 0$ being treated in Section 2.5.

2.2.2 Problem Statement

This study lies in the experimentally based premise that a true physical hysteretic element is BIBO stable, which means that, for any bounded input signal $x(t)$, the hysteretic response is also bounded. Thus the Bouc–Wen model Φ_{BW} should keep the BIBO stability property in order to be considered an adequate candidate to model real physical systems. Example 1 gives an example of a set of parameters A, β, γ, n such that, for a particular bounded input $x(t)$, the corresponding output $\Phi_{BW}(x)(t)$ given by the Bouc–Wen model (2.4)–(2.5) is unbounded. Thus, this set of parameters does not correspond to the description of a hysteretic physical element. This motivates the following problem:

> Given the parameters $0 < \alpha < 1$, $k > 0$, $D > 0$, A, β, γ with $\beta + \gamma \neq 0$ and $n > 1$, find the set of initial conditions $z(0)$ for which the Bouc–Wen model (2.4)–(2.5) is BIBO stable.

Note that when this set is empty, this means that the Bouc–Wen model is not BIBO stable. The solution to this problem will enable different sets of parameters and initial conditions to be classified and, additionally, to determine explicit bounds for the hysteretic variable $z(t)$.

2.2.3 Classification of the BIBO-Stable Bouc–Wen Models

The following set is introduced:

$$\Omega_{\alpha,k,D,A,\beta,\gamma,n} = \{z(0) \in R \text{ such that } \Phi_{BW} \text{ is BIBO stable for all } C^1 \text{ input signals } x(t) \text{ with fixed values of the parameters } \alpha, k, D, A, \beta, \gamma, n\} \quad (2.6)$$

If the set $\Omega_{\alpha,k,D,A,\beta,\gamma,n}$ is empty, then, for any initial condition $z(0)$, there exists a bounded signal $x(t)$ such that the corresponding hysteretic output $\Phi_{BW}(t)$ is unbounded; that is the set of parameters $\{\alpha, k, D, A, \beta, \gamma, n\}$ does not correspond to a BIBO-stable Bouc–Wen model. The emptiness of the set $\Omega_{\alpha,k,D,A,\beta,\gamma,n}$ is thus equivalent to the instability of the Bouc–Wen model, and for this reason the rest of the analysis is devoted to determining explicitly the set $\Omega_{\alpha,k,D,A,\beta,\gamma,n}$ as a function of the Bouc–Wen model parameters.

Let $z(0)$ be an element of $\Omega_{\alpha,k,D,A,\beta,\gamma,n}$. Then, for any bounded C^1 input $x(t)$, the output $\Phi_{BW}(x)(t)$ is bounded. This implies by

Equation (2.4) that the output $z(t)$ of the differential equation (2.5) should be bounded. This means that the set[1]

$$\Omega_{A,\beta,\gamma,n} = \{z(0) \in \mathbb{R} \text{ such that } z(t) \text{ is bounded for any } C^1 \text{ bounded input signal } x(t) \text{ with fixed values of the parameters } A, \beta, \gamma, n\} \tag{2.7}$$

is such that $\Omega_{\alpha,k,D,A,\beta,\gamma,n} \subset \Omega_{A,\beta,\gamma,n}$. The inclusion in the other way is immediate, which shows that $\Omega_{\alpha,k,D,A,\beta,\gamma,n} = \Omega_{A,\beta,\gamma,n}$. The importance of this equality stems for the fact that it is easier to determine the set $\Omega_{A,\beta,\gamma,n}$. Note that an empty set $\Omega_{A,\beta,\gamma,n}$ means that, with the chosen parameters A, β, γ, n, the Bouc–Wen model does not represent adequately the behaviour of a real hysteretic system Φ_s (see Example 1). The following set is also defined:

$$\Omega^{\star}_{A,\beta,\gamma,n} = \{z(0) \in \mathcal{R} \text{ such that } z(t) \text{ is bounded for any } C^1 \text{ input signal } x(t) \text{ with fixed values of the parameters } A, \beta, \gamma, n\} \tag{2.8}$$

Note that $\Omega^{\star}_{A,\beta,\gamma,n} \subset \Omega_{A,\beta,\gamma,n}$. With the notation introduced above, the main results of this section are given in the following theorem.

Theorem 1. *Let $x(t)$, $t \in [0, \infty)$ be a C^1 input signal and*

$$z_0 \triangleq \sqrt[n]{\frac{A}{\beta+\gamma}} \quad \text{and} \quad z_1 \triangleq \sqrt[n]{\frac{A}{\gamma-\beta}} \tag{2.9}$$

Then, the BIBO-stable Bouc–Wen models are identified in Table 2.1 Moreover,

$$\Omega^{\star}_{A,\beta,\gamma,n} = \Omega_{A,\beta,\gamma,n} \triangleq \Omega \tag{2.10}$$

Proof. First a check is made to see whether the differential equation (2.5) has a unique solution. Equation (2.5) may be seen as a nonautonomous locally Lipschitz system where the dependence on time is continuous. The local Lipschitz property is due to the fact

---

[1] The correct notation would be $\Omega_{A,\beta,\gamma,n,D}$. However, it will be seen later that this set does not depend on the parameter D.

Table 2.1 Classification of the BIBO-stable Bouc–Wen models

Case		Ω	Upper bound on $\lvert z(t)\rvert$	Class
$A > 0$	$\beta+\gamma > 0$ and $\beta-\gamma \geq 0$	$\mathbb{R}$	$\max(\lvert z(0)\rvert, z_0)$	I
	$\beta-\gamma < 0$ and $\beta \geq 0$	$[-z_1, z_1]$	$\max(\lvert z(0)\rvert, z_0)$	II
$A < 0$	$\beta-\gamma > 0$ and $\beta+\gamma \geq 0$	$\mathbb{R}$	$\max(\lvert z(0)\rvert, z_1)$	III
	$\beta+\gamma < 0$ and $\beta \geq 0$	$[-z_0, z_0]$	$\max(\lvert z(0)\rvert, z_1)$	IV
$A = 0$	$\beta+\gamma > 0$ and $\beta-\gamma \geq 0$	$\mathbb{R}$	$\lvert z(0)\rvert$	V
All other cases		$\emptyset$		

that $n > 1$. The time-dependent part of Equation (2.5) is the term $\dot{x}$, which is continuous as $x(t)$ has been assumed to be C^1. Thus, by Theorem 8 in the Appendix, a unique solution of Equation (2.5) does exist on some time interval $[0, t_0]$. Note that for $0 \leq n < 1$, the differential equation (2.5) does not verify the local Lipschitz property and thus the solutions may not be unique.

According to Table 2.1, the following three cases are considered: $A > 0$, $A < 0$ and $A = 0$. The proof is developed in detail only for the case $A > 0$. The other two cases can be treated in a similar way.

Assuming $A > 0$, consider the following four possibilities:

$$
\begin{aligned}
P_1 &: \ \beta+\gamma > 0 \text{ and } \beta-\gamma \geq 0 \\
P_{21} &: \ \beta+\gamma > 0,\ \beta-\gamma < 0 \text{ and } \beta \geq 0 \\
P_{22} &: \ \beta+\gamma > 0,\ \beta-\gamma < 0 \text{ and } \beta < 0 \\
P_3 &: \ \beta+\gamma < 0
\end{aligned}
$$

Note that the case P_{21} can be reduced to $\beta-\gamma < 0$ and $\beta \geq 0$ since $\beta+\gamma > 0$ is implied by the other two inequalities.

Case P_1

The Lyapunov function candidate $V(t) = z(t)^2/2$ is considered. Its derivative takes different forms depending on the signs of $\dot{x}$ and z. Then, consider the following sets:

$$Q_1 = \{\dot{x} \geq 0 \text{ and } z \geq 0\}$$

$$Q_2 = \{\dot{x} \geq 0 \text{ and } z \leq 0\}$$

$$Q_3 = \{\dot{x} \leq 0 \text{ and } z \geq 0\}$$

$$Q_4 = \{\dot{x} \leq 0 \text{ and } z \leq 0\}$$

Denoting $\dot{V}_{|Q_1}$ as the expression of the derivative of the function V over the set Q_1, then $\dot{V}_{|Q_1} = z\dot{x}D^{-1}[A-(\beta+\gamma)z^n]$. Thus $\dot{V}_{|Q_1} \leq 0$ for $z \geq z_0$. Also, $\dot{V}_{|Q_2} = z\dot{x}D^{-1}[A+(\beta-\gamma)|z|^n]$. In this case, $\dot{V}_{|Q_2} \leq 0$ for all values of z. The same conclusion is drawn in the case of Q_3, since $\dot{V}_{|Q_3} = z\dot{x}D^{-1}[A+(\beta-\gamma)z^n]$. Finally, $\dot{V}_{|Q_4} = z\dot{x}D^{-1}[A-(\beta+\gamma)|z|^n]$. Thus, $\dot{V}_{|Q_4} \leq 0$ for $|z| \geq z_0$.

It is then concluded that, for all the possibilities of the signs of $\dot{x}$ and z, then $\dot{V} \leq 0$ for all $|z| \geq z_0$. By Theorem 12 in the Appendix it is concluded that $z(t)$ is bounded for every continuous function $\dot{x}(t)$ and every initial condition $z(0)$. This means that $\Omega^{\star}_{A,\beta,\gamma,n} = \mathbb{R}$. Since $\Omega^{\star}_{A,\beta,\gamma,n} \subset \Omega_{A,\beta,\gamma,n}$, this implies that $\Omega^{\star}_{A,\beta,\gamma,n} = \Omega_{A,\beta,\gamma,n} = \mathbb{R}$. The bounds on $z(t)$ can be derived from Theorem 12 as follows:

1. If the initial condition of z is such that $|z(0)| \leq z_0$ then $|z(t)| \leq z_0$ for all $t \geq 0$.
2. If the initial condition of z is such that $|z(0)| \geq z_0$ then $|z(t)| \leq |z(0)|$ for all $t \geq 0$.

So far, class I of Table 2.1 has been identified.

Case P_{21}

Again, the derivative of $V(t)$ depends on the signs of $\dot{x}$ and z. Indeed, $\dot{V} \leq 0$ in the following regions:

$$\{\dot{x} \geq 0 \text{ and } z \geq 0 \text{ and } z \geq z_0\} \tag{2.11}$$

$$\{\dot{x} \geq 0 \text{ and } z \leq 0 \text{ and } |z| \leq z_1\} \tag{2.12}$$

$$\{\dot{x} \leq 0 \text{ and } z \geq 0 \text{ and } z \leq z_1\} \tag{2.13}$$

$$\{\dot{x} \leq 0 \text{ and } z \leq 0 \text{ and } |z| \geq z_0\} \tag{2.14}$$

The condition $\beta \geq 0$ leads to $z_1 \geq z_0$. From regions (2.11) to (2.14) it is concluded that $\dot{V} \leq 0$ for every $z_0 \leq |z| \leq z_1$ independently of the sign of $\dot{x}$. By Theorem 12 in the Appendix it is concluded that $z(t)$ is bounded for every continuous function $\dot{x}(t)$ and any initial state $z(0)$ such that $|z(0)| \leq z_1$. This means that $[-z_1, z_1] \subset \Omega^{\star}_{A,\beta,\gamma,n}$.

Now, take $z(0) \notin [-z_1, z_1]$; it is claimed that a bounded C^1 signal $x(t)$ exists such that the corresponding signal $z(t)$ is unbounded. The construction of such a signal is done in Lemma 1, which means that

$z(0) \notin \Omega_{A,\beta,\gamma,n}$. This implies that $\Omega_{A,\beta,\gamma,n} \subset [-z_1, z_1]$. Since $\Omega^{\star}_{A,\beta,\gamma,n} \subset \Omega_{A,\beta,\gamma,n}$, then $\Omega^{\star}_{A,\beta,\gamma,n} = \Omega_{A,\beta,\gamma,n} = [-z_1, z_1]$. Using Theorem 12 again, the following bound can be obtained: $|z(t)| \leq \max(|z(0)|, z_0)$.

Lemma 1. *Take $z(0) \notin [-z_1, z_1]$; then a bounded C^1 signal $x(t)$ exists such that the corresponding signal $z(t)$ is unbounded.*

Proof. Assume that $\beta - \gamma < 0$, $\beta \geq 0$ and assume that the initial condition $z(0)$ is such that $|z(0)| > z_1$. Take $z(0) > z_1$ (the construction is similar in the case $z(0) < -z_1$) and define the signal

$$\dot{x} = \frac{D}{A + (\beta - \gamma)z^n} \tag{2.15}$$

Since $z(0) > z_1 > 0$, then $A + (\beta - \gamma)z(0)^n < 0$, which means that the solution $z(t)$ of the differential equation (2.5) is well defined, at least during a maximal time interval $[0, t_1)$ in which $z(t) > z_1$. For $0 \leq t < t_1$, then $z > 0$ and $\dot{x} < 0$. Thus Equation (2.5) reduces to

$$\dot{z} = D^{-1}\dot{x}\,[A + (\beta - \gamma)\,z^n] \tag{2.16}$$

Combining Equations (2.15) and (2.16), it follows that

$$\dot{z} = 1 \tag{2.17}$$

Integrating Equation (2.17) gives, for $0 \leq t < t_1$,

$$z(t) = t + z(0) \tag{2.18}$$

$$\dot{x}(t) = \frac{D}{A + (\beta - \gamma)\,[t + z(0)]^n} \tag{2.19}$$

as the function $z(t)$ is increasing, the conditions of existence of $\dot{x}$ in Equations (2.15) and (2.19) are satisfied for any $t \geq 0$. This means that $t_1 = \infty$; that is $\dot{x}$ is well defined for all $t \geq 0$ and a solution of the differential equation (2.5) (given by Equation (2.18)) exists over $t \in \mathbb{R}_+$. From Equation (2.19), it follows that $\dot{x} \in L^1$ as $n > 1$. This implies by Lemma 13 (see the Appendix) that $x(t)$ goes to a finite limit as t goes to infinity, which means that $x(t)$ is bounded. Thus a bounded C^1 signal $x(t)$ has been constructed with an unbounded corresponding signal $z(t)$.

It has therefore been proved that, for $\beta-\gamma<0$ and $\beta\geq 0$, then $\Omega^{\star}_{A,\beta,\gamma,n}=\Omega_{A,\beta,\gamma,n}=[-z_1,z_1]$ and $|z(t)|\leq \max(|z(0)|,z_0)$. This means that class II of Table 2.1 has been identified.

Case P_{22}

For $\beta<0$ define

$$z_1<z_2\triangleq\sqrt[n]{\frac{A}{\gamma-\beta/2}}<z_0$$

and consider that the initial condition $z(0)\geq 0$ is such that $z(0)\leq z_2$ (the case $z(0)\leq 0$ can be treated in a similar way). Take $\dot{x}(t)=a$ for some positive constant a. Then, from Equation (2.5) and from the fact that $\beta<0$, then $z(t)\geq 0$ in a maximal time interval $[0,t_3]$ so that in $[0,t_3]$ it is found that

$$\dot{z}\geq aD^{-1}(A-\gamma|z|^n)\geq aD^{-1}(A-\gamma z_2^n)=aD^{-1}A\left(1-\frac{\gamma}{\gamma-\beta/2}\right)\triangleq b>0 \tag{2.20}$$

Equation (2.20) shows that $z(t)$ will increase and reach the value z_2 in a finite time t_2 and that $t_3\geq t_2$. At this point, the expression in Equation (2.15) with $\tau=t-t_2$ and $z(0)=z_2$ is chosen for $\dot{x}(\tau)$, which gives the conditions of Lemma 1. This means that for any initial condition $z(0)\leq z_2$, it is possible to construct a bounded C^1 signal $x(t)$ such that the corresponding signal $z(t)$ is unbounded. For initial conditions such that $z(0)\geq z_2$, then $z(0)>z_1$, which again gives the conditions of Lemma 1. Therefore the expression in Equation (2.15) is chosen for $\dot{x}(t)$.

It has thus been proved that, for $\beta<0$, then $\Omega_{A,\beta,\gamma,n}=\emptyset$. Since $\Omega^{\star}_{A,\beta,\gamma,n}\subset\Omega_{A,\beta,\gamma,n}$, this implies that $\Omega^{\star}_{A,\beta,\gamma,n}=\Omega_{A,\beta,\gamma,n}=\emptyset$.

Case P_3

Assume that $z(0)\geq 0$ (a similar analysis can be done for the case $z(0)\leq 0$) and define

$$k_1=-\frac{\beta+\gamma}{D},\qquad k_2=\frac{A}{\beta+\gamma}$$

By choosing $\dot{x} > 0$ it is found from Equation (2.5) that

$$\dot{z} = k_1 \dot{x} \left(k_2 + z^n\right) \tag{2.21}$$

On the other hand, since $n > 1$, the quantity

$$S = \int_{z(0)}^{\infty} \frac{\mathrm{d}u}{k_2 + u^n} > 0$$

is finite. Choosing for the signal $x(t)$ any increasing function such that $x(0) = 0$ and $\lim_{t\to\infty} x(t) = S/k_1$, it follows from Equation (2.21) that

$$\lim_{t\to\infty} \int_{z(0)}^{z(t)} \frac{\mathrm{d}u}{k_2 + u^n} = S \tag{2.22}$$

Equation (2.22) shows that $\lim_{t\to\infty} z(t) = \infty$, so a bounded signal $x(t)$ with an unbounded output $z(t)$ has been constructed for every initial condition $z(0)$. This means that $\Omega_{A,\beta,\gamma,n} = \emptyset$, which implies that $\Omega^{\star}_{A,\beta,\gamma,n} = \Omega_{A,\beta,\gamma,n} = \emptyset$.

The proof for the case with $A > 0$ is now concluded with the characterization of classes I and II. The cases $A < 0$ and $A = 0$ can be treated in a similar way to identify classes III to V, thus ending the proof of Theorem 1.

2.2.4 Practical Remarks

Table 2.1 shows that classes I to V are BIBO stable. A class is composed of a range for the Bouc–Wen model parameters and a range for the initial condition $z(0)$ of the hysteretic part of the model.

The fact that $\Omega^{\star}_{A,\beta,\gamma,n} = \Omega_{A,\beta,\gamma,n} \triangleq \Omega$ means that, for all classes, the boundedness of the hysteretic signal $z(t)$:

(a) depends only on the parameters A, γ, β and n;
(b) is independent of the boundedness of the input signal $x(t)$: for every input signal $x(t)$ (under the only assumption that it is C^1), the output $z(t)$ is always bounded if the set Ω is nonempty, and if $z(0) \in \Omega$.

Property (b) is particularly important for system control theory: indeed, when $x(t)$ is a closed-loop signal, it cannot be assumed a priori that it is bounded. This property of the Bouc–Wen model will be exploited in Chapter 6 to develop control strategies. Property (b) also shows that the solution $z(t)$ is defined for all $t \geq 0$ (see Theorem 10 in the Appendix).

Note that, for Class V, $z(t) = 0$ for all $t \geq 0$ if $z(0) = 0$. Since the Bouc–Wen model is often used with an initial condition $z(0) = 0$, this means that the class V corresponds, in this case, to a linear behaviour and is thus irrelevant from the point of view of the description of hysteretic systems.

Finally, it can be noted that, in all cases where a nonempty set Ω exists, the parameter β is nonnegative.

2.3 FREE MOTION OF A HYSTERETIC STRUCTURAL SYSTEM

Section 2.2 has analysed the stability properties of the Bouc–Wen model. It has shown that, for the Bouc–Wen model to be BIBO stable, it should belong to classes I to V of Table 2.1. Class V has been shown to be irrelevant as, in practice, it corresponds to the description of a linear behaviour. For this reason, only classes I to IV are considered in the present section.

2.3.1 Problem Statement

As a prototype system, a structural isolation scheme is considered, as illustrated in Figure 2.3. It is modelled as one-degree-of-freedom system with mass $m > 0$ and viscous damping $c > 0$ plus a restoring force Φ characterizing a hysteretic behaviour of the isolator material.

This system is described by the second-order differential equation

$$m\ddot{x} + c\dot{x} + \Phi(x)(t) = f(t) \tag{2.23}$$

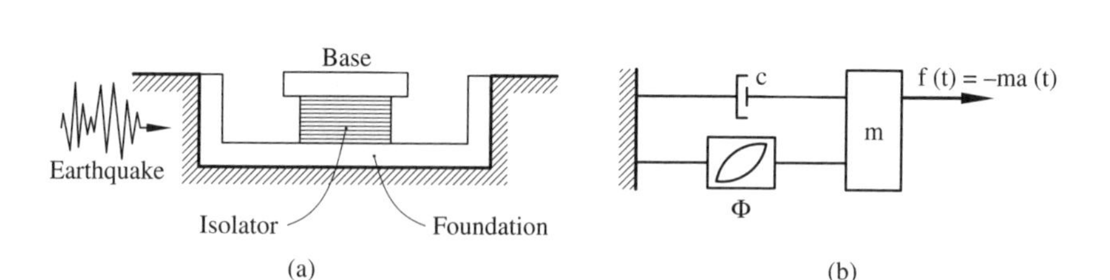

Figure 2.3 Base isolation device (a) and its physical model (b).

with initial conditions $x(0)$ and $\dot{x}(0)$ and excited by a force $f(t)$, like the one of the form $-ma(t)$ in the case of an earthquake with ground acceleration $a(t)$. The restoring force is assumed to be described by the Bouc–Wen model:

$$\Phi(x)(t) = \alpha kx(t) + (1-\alpha)Dkz(t) \tag{2.24}$$

$$\dot{z} = D^{-1}\left(A\dot{x} - \beta|\dot{x}||z|^{n-1}z - \gamma\dot{x}|z|^n\right) \tag{2.25}$$

where $n > 1$, $D > 0$, $k > 0$, $0 < \alpha < 1$ and $\beta + \gamma \neq 0$. The purpose of this section is to study the free motion of system (2.23)–(2.25), that is with $f(t) = 0$, in order to analyse its asymptotic trajectories.

In real applications, the base-isolation devices are designed to dissipate the energy introduced in the structure by external perturbations. In the absence of disturbances, the structure is in free motion so that, when its initial conditions are not zero, the structure dissipates the energy due to the initial conditions and goes to rest asymptotically. Base-isolation devices have been often modelled by the Bouc–Wen model (see Chapter 1 for detailed references), so this model has to reproduce some general physical properties of these devices independently of the exciting input. In particular, for the Bouc–Wen based model (2.23)–(2.25) that is supposed to reproduce the behaviour of a base-isolation device, it is desirable that the velocity $\dot{x}$ of the mass goes asymptotically to zero and that its position x goes to a constant. The next section shows that this is the case for classes I and II.

2.3.2 Asymptotic Trajectories

In this section, the asymptotic behaviour is analysed of the system defined by Equations (2.23) to (2.25) in the absence of an external excitation. The main result of this section is given in the following theorem.

Theorem 2. *For every initial conditions $x(0) \in \mathbb{R}$, $\dot{x}(0) \in \mathbb{R}$ and $z(0) \in \Omega \neq \emptyset$, the following holds:*

(*a*) *For all classes I to IV of Table 2.1, the signals $x(t)$, $\dot{x}(t)$ and $z(t)$ are bounded and C^1.*

(*b*) *Assume that the Bouc–Wen model belongs to classes I or II. Then, constants x_∞ and z_∞ exist that depend on the Bouc–Wen model parameters $(\alpha, D, k, A, \beta, \gamma, n)$, the system parameters (m, c) and the initial conditions $(x(0), \dot{x}(0), z(0))$, and a constant $\bar{c}$ exists that* *depends on the parameters* m, k, A, α, β, γ *such that, for all* $c \geq \bar{c}$,

$$\lim_{t\to\infty} x(t) = x_\infty \tag{2.26}$$

$$\lim_{t\to\infty} z(t) = z_\infty \tag{2.27}$$

$$\alpha x_\infty + (1-\alpha) D z_\infty = 0 \tag{2.28}$$

Furthermore,

$$\dot{x} \in L^1([0,\infty)) \qquad \text{and} \qquad \lim_{t\to\infty} \dot{x}(t) = 0 \tag{2.29}$$

Proof. First, part (a) is addressed. A state–space system realization of Equations (2.23) to (2.25) is

$$\dot{x}_1 = x_2 \tag{2.30}$$

$$\dot{x}_2 = m^{-1}[-c x_2 - \alpha k x_1 - (1-\alpha) k D z] \tag{2.31}$$

$$\dot{z} = D^{-1}\left(A x_2 - \beta |x_2|\, |z|^{n-1} z - \gamma x_2 |z|^n\right) \tag{2.32}$$

where $x_1 = x$. Since Equations (2.30) to (2.32) is locally Lipschitz, then a C^1 solution $(x_1(t), x_2(t), z(t))$ exists over some time interval $[0, t_0)$. It has been seen (Theorem 1 in Section 2.2.3) that $z(t)$ given by Equation (2.25) is bounded for every C^1 signal x (bounded or not) once $z(0)$ belongs to $\Omega \neq \emptyset$. Thus Equation (2.23) can be written in the form

$$m\ddot{x} + c\dot{x} + \alpha k x = -(1-\alpha) D k z \tag{2.33}$$

which may be seen as an exponentially stable second-order linear system excited by a bounded external input signal $-(1-\alpha)Dkz$. This implies that the signals $x_1(t)$, $x_2(t)$ and $z(t)$ are bounded (by Lemma 12 in the Appendix) and thus $t_0 = \infty$. This proves part (a) of Theorem 2.

Let us move to part (b) of Theorem 2 and proceed in two steps:

1. It will be shown that $\dot{x}$ belongs to L^1.
2. This property will be used to complete the proof of the theorem.

Proof of step 1. The following two cases are considered:

P_1: $|z(t)| > z_0$ for all $t \geq 0$
P_2: There exists some $t_0 < \infty$ such that $|z(t_0)| \leq z_0$

The case P_1 is treated first. Using the results of Section 2.2 it follows that the time function $z(t)^2$ is nonincreasing. Since it is bounded, it goes to a limit $z_\infty^2 \geq z_0^2$. Consider the case where $z(0) > 0$ (the analysis is similar in the case $z(0) < 0$). By continuity of z, then $z(t) \geq z_\infty \geq z_0 > 0$ for all $t \geq 0$. Take $\varepsilon > 0$; then some $t_1 < \infty$ exists such that

$$z_\infty^n \leq z(t)^n \leq z_\infty^n + \varepsilon \qquad \text{for all} \quad t \geq t_1 \tag{2.34}$$

Multiplying by x_2 and integrating both parts of Equation (2.34), the following is obtained for any $T \geq 0$:

$$\begin{aligned} z_\infty^n \int_{t_1}^{t_1+T} x_2(t)\mathrm{d}t - \varepsilon \int_{t_1}^{t_1+T} |x_2(t)|\, \mathrm{d}t &\leq \int_{t_1}^{t_1+T} x_2(t)z(t)^n \mathrm{d}t \\ &\leq z_\infty^n \int_{t_1}^{t_1+T} x_2(t)\mathrm{d}t + \varepsilon \int_{t_1}^{t_1+T} |x_2(t)| \end{aligned} \tag{2.35}$$

On the other hand, using the fact that $\beta \geq 0$ (see Section 2.2.4), the following is obtained from Equations (2.32) and (2.34):

$$\beta\, |x_2|\, z_\infty^n \leq \beta\, |x_2|\, z^n = -D\dot{z} + Ax_2 - \gamma x_2 z^n \tag{2.36}$$

Here two subcases need to be discussed:

P_{11}: $\beta > 0$
P_{12}: $\beta = 0$

Let us focus first on the subcase P_{11}. Integrating both parts of inequality (2.36) and using Equation (2.35), it follows that

$$\begin{aligned} \int_{t_1}^{t_1+T} |x_2(t)|\mathrm{d}t \leq &-\frac{D}{\beta z_\infty^n}\, [z(t_1+T) - z(t_1)] \\ &+ \left(\frac{A}{\beta z_\infty^n} - \frac{\gamma}{\beta}\right)\, [x_1(t_1+T) - x_1(t_1)] \\ &+ \frac{|\gamma|\varepsilon}{\beta z_\infty^n} \int_{t_1}^{t_1+T} |x_2(t)|\mathrm{d}t \end{aligned} \tag{2.37}$$

If $\gamma = 0$, then using part (a) of Theorem 2, it follows from Equation (2.37) that $x_2 \in L^1$ as T is arbitrary. If $\gamma \neq 0$, choosing $\varepsilon = \beta z_\infty^n/(2|\gamma|)$ in Equation (2.37) shows that $x_2 \in L^1$.

The subcase P_{12} is now considered. Note that the parameter β can be zero only for class II. In this case, Table 2.1 shows that $\gamma > 0$. By assumption P_1, $z(t) > z_0$ for all $t \geq 0$; thus it follows that $z^n > z_0^n = A/\gamma$ or equivalently $A - \gamma z^n < 0$.

Using Equation (2.32) with $\beta = 0$ leads to

$$x_2 = \frac{D\dot{z}}{A - \gamma z^n}$$

Since z is nonincreasing, $\dot{z} \leq 0$, which implies that $x_2(t) \geq 0$ for all $t \geq 0$. Then

$$\int_0^t |x_2(\tau)|\,\mathrm{d}\tau = \int_0^t x_2(\tau)\mathrm{d}\tau = x_1(t) - x_1(0)$$

which, using part (a) of Theorem 2, shows that $x_2 \in L^1$.

Case P_2 will now be treated. Taking the derivative of Equation (2.31) gives

$$\lambda_2 \ddot{x}_2 + \lambda_1 \dot{x}_2 + \lambda_0 x_2 = -\dot{z} \tag{2.38}$$

where

$$\lambda_2 = \frac{m}{(1-\alpha)Dk}, \qquad \lambda_1 = \frac{c}{(1-\alpha)Dk}, \qquad \lambda_0 = \frac{\alpha}{(1-\alpha)D} \tag{2.39}$$

By assumption P_2, $|z(t_0)| \leq z_0$. Using the bound on $|z(t)|$ of Table 2.1 for $A > 0$, it follows that $|z(t)| \leq z_0$ for all $t \geq t_0$. From Equation (2.32), for all $t \geq t_0$,

$$0 \leq D\dot{z}x_2 = [A - (\beta+\gamma)\,|z|^n]\,x_2^2 \leq Ax_2^2 \qquad \text{for } zx_2 \geq 0$$

$$Ax_2^2 \leq D\dot{z}x_2 = [A + (\beta-\gamma)\,|z|^n]\,x_2^2 \leq \frac{2A\beta}{\beta+\gamma}x_2^2 \qquad \text{for } zx_2 \leq 0$$

(class I Bouc–Wen model)

$$\frac{2A\beta}{\beta+\gamma}x_2^2 \leq D\dot{z}x_2 = [A + (\beta-\gamma)\,|z|^n]\,x_2^2 \leq Ax_2^2 \qquad \text{for } zx_2 \leq 0$$

(class II Bouc–Wen model)

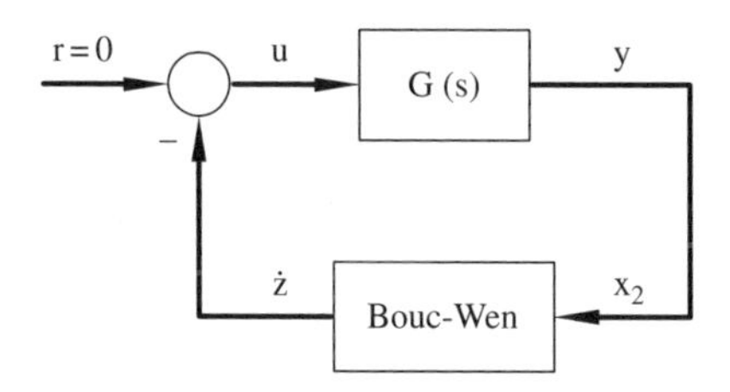

Figure 2.4 Equivalent description of system (2.30)–(2.32).

Consequently, the following holds in all cases:

$$0 \leq \dot{z}x_2 \leq \delta x_2^2 \tag{2.40}$$

where $\delta = (A/D) \max([1, 2\beta/(\beta+\gamma])$.

In order to conclude the proof of case P_2, stability issues presented in the Appendix (Sections A.3.3 and A.3.4) are invoked.

First note that the system (2.30)–(2.32) can be viewed under the feedback connection illustrated in Figure 2.4, where the reference signal is $r = 0$, the input signal is $u = -\dot{z}$, the output signal is $y = x_2$ and the transfer function $G(s)$ is given by

$$G(s) = \frac{1}{\lambda_2 s^2 + \lambda_1 s + \lambda_0}$$

By Equation (2.40), the nonlinearity that represents the Bouc–Wen model belongs to the sector $[0, \delta]$ (see Section A.3.3 in the Appendix). The idea is to use this fact to prove that the feedback connection is such that the state $(x_2, \dot{x}_2)$ goes exponentially to zero using Theorem 13 in the Appendix. It is to be noted that the Bouc–Wen nonlinearity is not memoryless in this case. However, since it is known that z is bounded, it can be checked easily that the stability proof of the feedback connection is exactly the same as in Theorem 13. In the following, the simplified version of Theorem 13 given in Theorem 14 is used. Then the conditions under which $\text{Re}[1 + \delta G(\text{j}\omega)] > 0$ for all $\omega \in (-\infty, \infty)$ need to be checked. This is equivalent to checking the conditions under which

$$f(\eta) = \lambda_2^2\eta^2 + \eta\left(-2\lambda_0\lambda_2 + \lambda_1^2 - \delta\lambda_2\right) + \lambda_0^2 + \delta\lambda_0 > 0 \quad \text{for all } \eta = \omega^2 \tag{2.41}$$

Equation (2.41) is a second-order algebraic equation in the variable η. If its discriminant Δ is negative, then the function $f(\eta)$ is always positive. If $\Delta \geq 0$, then $f(\eta) = 0$ has real roots η_1 and η_2 that have the same sign. If $\eta_1 < 0$ and $\eta_2 < 0$, then $f(\eta) > 0$ for all values of $\eta = \omega^2 \geq 0$. The only case where $f(\eta) < 0$ with $\eta \geq 0$ occurs when the sum of roots is positive. This condition can be written as

$$c < \sqrt{mk\left[A(1-\alpha)\max\left(1, \frac{2\beta}{\beta+\gamma}\right) + 2\alpha\right]} \triangleq \bar{c} \tag{2.42}$$

Thus, it has finally been proved that $c \geq \bar{c}$ implies that $\mathrm{Re}[1 + \delta G(\mathrm{j}\omega)] > 0$ for all $\omega \in (-\infty, \infty)$. Then, using Theorem 13, it follows that the state x_2 goes exponentially to zero. This implies that $x_2 = \dot{x} \in L^1$. It has therefore been shown that, in both cases P_1 and P_2, $\dot{x} \in L^1$, thus concluding step 1 of the proof.

Proof of step 2. The fact that $x_2 \in L^1$, along with Equation (2.30), shows that $x_1 = x$ goes to a finite limit (Lemma 13 in the Appendix), which establishes Equation (2.26). Since x_2 is bounded and $\dot{x}_2$ is bounded (by Equation (2.31) and $x_2 \in L^1$, then the Barbalat lemma (Lemma 14 in the Appendix) can be used to ensure that

$$\lim_{t\to\infty} x_2(t) = 0$$

Now, taking the derivative of Equation (2.31) gives

$$m\ddot{x}_2 + c\dot{x}_2 + \alpha k x_2 = -(1-\alpha)kD\dot{z} \tag{2.43}$$

which may be seen as a stable second-order system excited by the input $-(1-\alpha)kD\dot{z}$.

Since $x_2 \in L^1$, it follows that $\dot{z} \in L^1 \bigcap L^\infty$ by using part (a) of Theorem 2 and Equation (2.32). From Equation (2.43) it is then concluded that $\dot{x}_2 \in L^1$ (Lemma 12 in the Appendix). Thus $\dot{x}_2$ is bounded (by Equation (2.31), $\ddot{x}_2$ is bounded (by Equation (2.43) and $\dot{x}_2 \in L^1$. By application of Barbalat's lemma, it follows that

$$\lim_{t\to\infty} \dot{x}_2(t) = 0$$

Using this property in Equation (2.31) along with Equation (2.29) gives

$$\lim_{t\to\infty} [\alpha k x_1(t) + (1-\alpha)kDz(t)] = 0 \tag{2.44}$$

Since x_1 goes to a finite limit, the expression (2.44) establishes Equations (2.27) and (2.28). This completes the proof of Theorem 2.

2.3.3 Practical Remarks

Theorem 2 shows that, for classes I and II of Table 2.1, the states $x(t)$ and $z(t)$ go asymptotically to constant values and that the velocity $\dot{x}$ goes to zero. This means that both classes are good candidates for the description of the real physical behaviour of a base-isolation device. Crucial to the proof of the theorem is the fact that $\dot{x} \in L^1$. It is to be noted that the condition $c \geq \bar{c}$ is only a sufficient condition for the validity of Theorem 2. Numerical simulations show that this condition is not necessary.

Theorem 2 demonstrates that the Bouc–Wen based model (2.23)–(2.25) behaves in accordance with observed experiments for real base-isolation devices when the model belongs to classes I and II. This is not the case for classes III and IV. Indeed, consider class III Bouc–Wen based model (2.23)–(2.25) given by the following parameters:

$$\alpha = 0.5, k = 2, D = 1, A = -2, \beta = 1, \gamma = 0, n = 1.1, m = 1, c = 1$$

Figure 2.5 gives the solution of the differential equations (2.23) to (2.25) with initial conditions $x(0) = 0$, $\dot{x}(0) = 0.1$ and $z(0) = 0$. It is observed that, after a transient, a limit cycle occurs. In fact, it can be checked using numerical simulations that, for a large number of values of $A < 0$ (that is for classes III and IV) and for arbitrary small initial conditions, limit cycles are observed. A mathematical proof of this property is difficult. Such a behaviour has not been observed for real systems like base-isolation devices. For this reason, it is considered that the negative values of the parameter A do not correspond to a physical behaviour of the Bouc–Wen model.

The conclusion that can be drawn from this section is that, from classes I to IV of Table 2.1, only the two classes I and II are relevant from the point of view of a description of physical phenomena. The

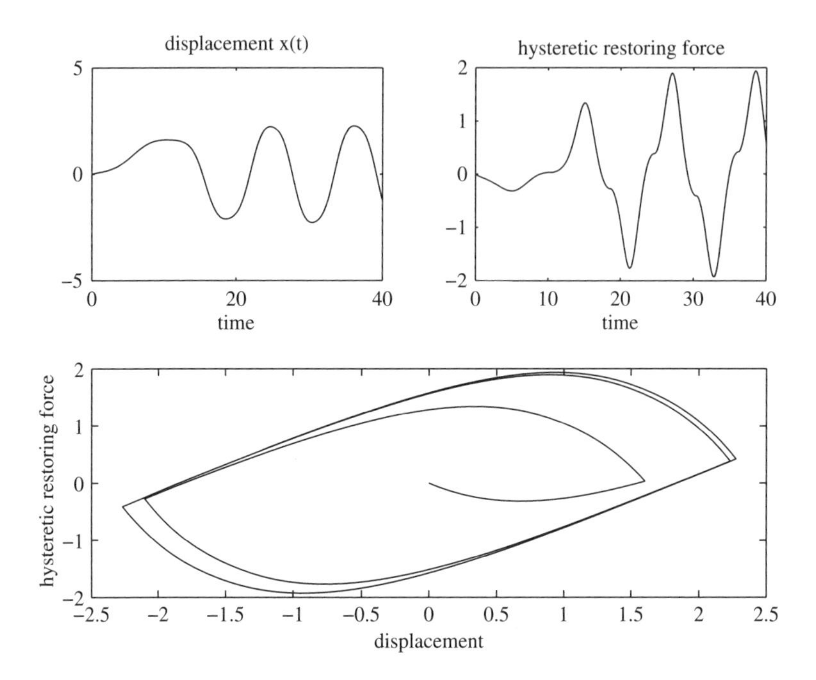

Figure 2.5 Limit cycles for a class III Bouc–Wen model.

rest of the classes of Table 2.1, that is classes III, IV and V, are irrelevant in practice.

2.4 PASSIVITY OF THE BOUC–WEN MODEL

In this section, the class I Bouc–Wen model is shown to be passive. In electric networks, passivity means that the network contains only passive elements so that the network does not generate energy. In mechanics, passivity is also related to energy dissipation (see Section A.3.1 in the Appendix). The Bouc–Wen model has been used mostly to describe passive devices in civil and mechanical engineering. This implies that this model has to be passive in order to represent adequately the physical elements that it describes.

The model (2.4)–(2.5) can be written in the form

$$\dot{x} = u \tag{2.45}$$

$$\dot{z} = D^{-1}\left(Au - \beta|u|\;|z|^{n-1}z - \gamma u|z|^{n}\right) \tag{2.46}$$

$$y = \alpha k x + (1-\alpha) D k z \tag{2.47}$$

The Bouc–Wen model is seen as a nonlinear system whose input is the velocity $\dot{x}(t) = u(t)$ and whose output is $y(t) = \Phi_{\mathrm{BW}}(x)(t)$. The displacement $x(t)$ and the variable $z(t)$ are seen as state variables. Combining Equation (2.46) along with the definition of class I (Table 2.1) results in

$$D\dot{z}z = Az\dot{x} - \beta|\dot{x}|\,|z|^{n+1} - \gamma\dot{x}|z|^{n}\,z \leq Az\dot{x} + (|\gamma| - \beta)\,|\dot{x}|\,|z|^{n+1} \leq Az\dot{x} \tag{2.48}$$

On the other hand, from Equations (2.45) and (2.47),

$$z\dot{x} = zu = \frac{y - \alpha k x}{(1-\alpha)Dk} u \tag{2.49}$$

Using Equations (2.48) and (2.49) gives

$$yu \geq 2l_1 z\dot{z} + 2l_2 x\dot{x} = \dot{W} \tag{2.50}$$

where

$$\begin{aligned} l_1 &= \frac{(1-\alpha)D^2 k}{2A} > 0 \\ l_2 &= \frac{\alpha k}{2} > 0 \\ W(x, z) &= l_1 z^2 + l_2 x^2 \end{aligned} \tag{2.51}$$

Equations (2.50) and (2.51) show that the Bouc–Wen model is passive with respect to the storage function $W(x, z)$.

2.5 LIMIT CASES

In this section the following limit cases are analysed for the Bouc–Wen model parameters: $n = 1$, $\alpha = 0$, $\alpha = 1$, $\beta + \gamma = 0$.

2.5.1 The Limit Case $n = 1$

The differential equation (2.5) remains locally Lipschitz for $n = 1$. However, the signal $\dot{x}$ constructed in Lemma 1 is no longer in L^1

so that the result of that lemma does not hold. This means that in Table 2.1 the expressions given for Ω are only subsets of the whole set Ω. With this observation, Table 2.1 holds also for the case $n = 1$. For example, for class II, a subset of Ω is given by $[-z_1, z_1]$ and for $z(0) \in [-z_1, z_1]$, an upper bound on $|z(t)|$ is given by $\max(|z(0)|, z_0)$, as indicated in Table 2.1. Theorem 2 holds for $n = 1$ with the only modification that $z(0)$ should belong to the subset of Ω given by Table 2.1 (and not to the whole set Ω).

2.5.2 The Limit Case $\alpha = 1$

For $\alpha = 1$ the hysteretic part in Equation (2.4) is zero so that the system (2.4) and (2.5) is linear and thus does not represent a hysteretic nonlinearity.

2.5.3 The Limit Case $\alpha = 0$

Table 2.1 holds for the case $\alpha = 0$. However, Theorem 2 does not hold necessarily as the linear system $m\ddot{x} + c\dot{x} + \alpha kx = 0$, which appears in Equation (2.33), is not exponentially stable and thus is not BIBO stable. This implies that the bounded input $-(1-\alpha)Dkz$ may not give rise to a bounded output x.

2.5.4 The Limit Case $\beta + \gamma = 0$

In this case, the upper bound on the variable $z(t)$ may depend on the input $x(t)$. Thus the Bouc–Wen model loses interest for control purposes as there is no longer a hysteretic output $z(t)$ that is bounded irrespective of the boundedness of the input $x(t)$.

2.6 CONCLUSION

This chapter has presented a classification of the possible Bouc–Wen models in terms of their bounded input–bounded output (BIBO) stability properties. It has been shown that only five classes I to V of Bouc–Wen models are BIBO stable. One of them (class V) has been shown to be irrelevant in practice since the hysteretic part of the model remains equal to zero when the initial condition is $z(0) = 0$.

The chapter has also analysed the asymptotic behaviour of a one-degree-of-freedom mechanical (structural) system with a hysteretic restoring force represented by the Bouc–Wen model. It has been shown that, for all classes, the displacement of the mass and its velocity are bounded. Furthermore, for classes I and II, the displacement of the mass goes asymptotically to a constant, the restoring force goes to zero and the velocity of the mass is in L^1 and goes to zero asymptotically. This behaviour is in accordance with experimental observations for real structures. On the other hand, numerical simulations have shown that classes III and IV do not describe adequately the behavior of real structures as the solution of the related differential equations lead to limit cycles that have not been observed experimentally.

Finally, it has been shown that class I is passive. This passivity property is related to the energy dissipation observed experimentally in devices described by the Bouc–Wen model. In a parallel work [8], the study of the thermodynamic admissibility of the Bouc–Wen model within the context of the endochronic theory led to the following result: the conditions $A > 0$ and $-\beta \leq \gamma \leq \beta$ are necessary and sufficient for the thermodynamic admissibility of the Bouc–Wen model. This means that classes II to V are not consistent with the laws of thermodynamics, while class I is consistent. Hence, class I is the only one that is BIBO stable, is compatible with the free motion of the real systems described by the Bouc–Wen model, is passive and is compatible with the laws of thermodynamics. For this reason, class I is dealt with exclusively in the rest of the book.

3

Forced Limit Cycle Characterization of the Bouc–Wen Model

3.1 INTRODUCTION

In this chapter, the input signal $x(t)$ is considered to be periodic with a loading–unloading shape. Such a signal is often used for identification purposes. The aim of the chapter is to show that the hysteretic output of the Bouc–Wen model goes asymptotically to a periodic solution and to give the analytic expression of this solution. More precisely, the chapter treats the following points:

1. It proves analytically that the response of the Bouc–Wen model to a class of T-periodic inputs is asymptotically T-periodic. Indeed, the fact that periodic inputs lead to periodic outputs (asymptotically) is well established for stable linear systems. However, for nonlinear systems this is not always the case; for example a Duffing oscillator may present chaotic behaviour as a response to a sine wave [124, page 614]. A good theorem that shows the existence of periodic solutions to periodic excitations in the general context of nondifferentiable systems (which is the case of the Bouc-Wen model) is given in Theorem 3 of Reference [125, page 148]. However, this theorem necessitates knowledge of a Lyapunov-like function, which is very difficult to find in the

Systems with Hysteresis: Analysis, Identification and Control using the Bouc–Wen Model
F. Ikhouane and J. Rodellar

context of nonlinear systems, and for this reason this theorem has not been used in this chapter.

2. The chapter gives an exact and explicit analytic expression of the limit cycle. Indeed, even though Theorem 3 of Reference [125] demonstrates the existence of periodic solutions to periodic inputs, it does not give an explicit analytic expression of the limit cycle. The exact description of limit cycles for nonlinear systems remains largely an open problem, even though nonrigorous approximate methods do exist (the harmonic balance principle for example) along with the well-known Poincaré maps [124]. The last technique requires some knowledge of the geometric structure of the phase space of the differential equation, knowledge that is lacking in the current literature devoted to the Bouc–Wen model and thus inhibiting the use of the Poincaré maps method in the present case.

The analytic expression of the limit cycle that is derived in this chapter will be used in Chapter 4 to determine the way in which the Bouc–Wen model parameters shape the limit cycle and in Chapter 5 to obtain an identification method for the Bouc–Wen model parameters.

3.2 PROBLEM STATEMENT

3.2.1 The Class of Inputs

In this chapter the input signal $x(t)$ is considered to be continuous on the time interval $[0, +\infty)$ and periodic of period $T > 0$. Furthermore, a specific structure is assumed for this function, which is illustrated in Figure 3.1 and detailed as follows:

- A scalar $0 < T^+ < T$ exists such that the signal x is C^1 on both intervals $(0, T^+)$ and (T^+, T) with the time derivative $\dot{x}(\tau) > 0$ for $\tau \in (0, T^+)$ and $\dot{x}(\tau) < 0$ for $\tau \in (T^+, T)$.
- $X_{\min} = x(0)$ and $X_{\max} = x(T^+) > X_{\min}$ are denoted as the minimal and maximal values of the input signal respectively.
- For a given integer m, the following time instants and intervals are defined:

$$t_m = mT \quad \text{and} \quad t_m^+ = t_m + T^+$$

$$I_m^+ = [t_m, t_m^+] \quad \text{and} \quad I_m^- = [t_m^+, t_{m+1}]$$

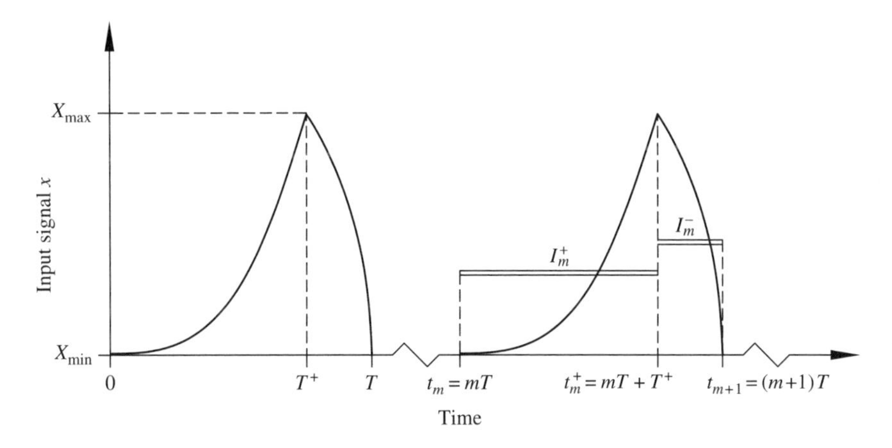

Figure 3.1 Illustration of the notation related to the input signal.

This class of inputs is very common in identification procedures for hysteretic systems [31] and include sine waves with or without offset, triangular inputs, etc. Due to the particular shape of this type of signal, they will be said to be *wave T-periodic.*

3.2.2 Problem Statement

Consider a T-periodic input signal $x(t)$ given as in Section 3.2.1 and the Bouc–Wen model (2.4)–(2.5). The problem under study is stated as follows:

1. Show that the hysteretic output converges asymptotically to a T-periodic solution.
2. Give an explicit analytic description of the periodic solution.

3.3 THE NORMALIZED BOUC–WEN MODEL

This section introduces a new form of the Bouc–Wen model: the normalized one. The motivation of this form is due to the following lemma.

Lemma 2. *Consider two Bouc–Wen models (2.4)–(2.5) whose parameters are such that*

$$n_2 = n_1 = n, \;\; A_2 = A_1, \;\; \beta_2 = \nu^n \beta_1, \;\; \gamma_2 = \nu^n \gamma_1,$$
$$D_2 = \nu D_1, \;\; \alpha_2 = \alpha_1, \;\; k_2 = k_1$$

where ν is a positive constant, and with the initial conditions $z_2(0) = z_1(0) = 0$. Then both models belong to the same class, and for any input signal $x(t)$ they deliver exactly the same output $\Phi_{BW}(x)(t)$.

Proof. The fact that both models belong to the same class follows directly from Table 2.1. Now consider a C^1 input signal $x(t)$. Then the corresponding hysteretic outputs are given by

$$\Phi_{BW1}(x)(t) = \alpha_1 k_1 x(t) + (1 - \alpha_1) D_1 k_1 z_1(t) \tag{3.1}$$
$$\dot{z}_1 = D_1^{-1} \left(A_1 \dot{x} - \beta_1 |\dot{x}| \, |z_1|^{n-1} z_1 - \gamma_1 \dot{x} |z_1|^n \right) \tag{3.2}$$

and

$$\Phi_{BW2}(x)(t) = \alpha_2 k_2 x(t) + (1 - \alpha_2) D_2 k_2 z_2(t) \tag{3.3}$$
$$\dot{z}_2 = D_2^{-1} \left(A_2 \dot{x} - \beta_2 |\dot{x}| \, |z_2|^{n-1} z_2 - \gamma_2 \dot{x} |z_2|^n \right) \tag{3.4}$$

From Equation (3.4), it follows that

$$\dot{z}_2 = \nu^{-1} D_1^{-1} \left(A_1 \dot{x} - \nu^n \beta_1 |\dot{x}| \, |z_2|^{n-1} z_2 - \nu^n \gamma_1 \dot{x} |z_2|^n \right) \tag{3.5}$$

Introducing the variable $z_\nu(t) = \nu z_2(t)$, Equation (3.5) gives

$$\dot{z}_\nu = D_1^{-1} \left(A_1 \dot{x} - \beta_1 |\dot{x}| \, |z_\nu|^{n-1} z_\nu - \gamma_1 \dot{x} |z_\nu|^n \right) \tag{3.6}$$

Note that Equations (3.2) and (3.6) are exactly the same. Since they have the same initial state ($z_\nu(0) = z_1(0) = 0$), then $z_\nu(t) = z_1(t)$ for all $t \geq 0$ due to the uniqueness of the solutions. Using the fact that $z_\nu(t) = \nu z_2(t)$, it follows that $z_2(t) = \nu^{-1} z_1(t)$. Combining this equality with Equation (3.3), it follows that

$$\Phi_{BW2}(x)(t) = \alpha_1 k_1 x(t) + (1 - \alpha_1)(\nu D_1) k_1 (\nu^{-1} z_1)(t) \tag{3.7}$$

Equations (3.1) and (3.7) show that $\Phi_{BW2}(x)(t) = \Phi_{BW1}(x)(t)$ for all $t \geq 0$, which completes the proof.

Lemma 2 means that the input–output behaviour of a Bouc–Wen model is not described by a unique set of parameters $\{\alpha, k, D, A, \beta, \gamma, n\}$. A drawback to this property is that identification procedures that use input–output data cannot determine the parameters of the Bouc–Wen model. To cope with this problem, users of the Bouc–Wen model often fix some parameters to arbitrary values. For example, in Reference [31] the coefficient $(1-\alpha)Dk$ of $z(t)$ in Equation (2.4) has been set to one and the parameter D has also been set to one. Other authors compare the shape of the limit cycle instead of comparing the identified parameters with their true values, as in Reference [92]. This fact makes it difficult to compare results of different identification methods by comparing the identified parameters. Thus it is necessary to elaborate some equivalent 'normalized' model whose parameters define in a unique way the input–output behaviour of the model, allowing a parametric-based comparison of identification methods for this hysteretic model. To this end, define

$$w(t) = \frac{z(t)}{z_0} \tag{3.8}$$

so that the model (2.4)–(2.5) can be written in the form

$$\Phi_{\mathrm{BW}}(x)(t) = \kappa_x x(t) + \kappa_w w(t) \tag{3.9}$$

$$\dot{w}(t) = \rho\left(\dot{x} - \sigma|\dot{x}(t)|\,|w(t)|^{n-1}w(t) + (\sigma-1)\dot{x}(t)|w(t)|^{n}\right) \tag{3.10}$$

where

$$\rho = \frac{A}{Dz_0} > 0, \quad \sigma = \frac{\beta}{\beta+\gamma} \geq 0,$$
$$\kappa_x = \alpha k > 0, \quad \kappa_w = (1-\alpha)Dkz_0 > 0 \tag{3.11}$$

Equations (3.9) and (3.10) define the so-called normalized form of the Bouc–Wen model. Note that if the initial condition $w(0)$ is such that $|w(0)| \leq 1$ then, by Theorem 1, $|w(t)| \leq 1$ for all $t \geq 0$. This means that the variable $z(t)$ has been scaled to unity. The fact that the normalized form of the Bouc–Wen model defines a bijective relationship between the input–output behaviour of the model and its parameters is demonstrated in Chapter 5. The normalized form has the advantage of having only five parameters to identify instead of

Table 3.1 Classification of the BIBO, passive and thermodynamically consistent normalized Bouc–Wen models

Case	Ω	Upper bound on $\|w(t)\|$	Class
$\sigma \geq \frac{1}{2}$	$\mathbb{R}$	$\max(\|w(0)\|, 1)$	I

the seven parameters for the standard form. Note that the normalized form of the Bouc–Wen model is exactly equivalent to its standard form. Indeed, for any input $x(t)$, both forms deliver exactly the same output $\Phi_{\mathrm{BW}}(t)$ taking into account that $w(0) = z(0)/z_0$.

The classification of the normalized Bouc–Wen models is given in Table 3.1. It can be seen that a single parameter σ is needed for this classification. Using this notation the following can be obtained from Equation (3.10):

$$\text{For } w(t) \geq 0 \text{ and } \dot{x}(t) \geq 0: \quad \dot{w}(t) = \rho\,(1 - w(t)^n)\,\dot{x}(t) \tag{3.12}$$

$$\text{For } w(t) \leq 0 \text{ and } \dot{x}(t) \geq 0: \quad \dot{w}(t) = \rho\,(1 + (2\sigma - 1)\,[-w(t)]^n)\,\dot{x}(t) \tag{3.13}$$

$$\text{For } w(t) \geq 0 \text{ and } \dot{x}(t) \leq 0: \quad \dot{w}(t) = \rho\,(1 + (2\sigma - 1) w(t)^n)\,\dot{x}(t) \tag{3.14}$$

$$\text{For } w(t) \leq 0 \text{ and } \dot{x}(t) \leq 0: \quad \dot{w}(t) = \rho\,(1 - [-w(t)]^n)\,\dot{x}(t) \tag{3.15}$$

For notational convenience, $w(t_m) = w_m$ and $w(t_m^+) = w_m^+$.

3.4 INSTRUMENTAL FUNCTIONS

In this section the case $\sigma \geq \frac{1}{2}$ is considered. The following functions $\varphi_{\sigma,n}^-$, $\varphi_{\sigma,n}^+$ and $\varphi_{\sigma,n}$ are defined, which will be helpful for the integration of Equation (3.10):

$$\varphi_{\sigma,n}^-(w) = \int_0^w \frac{\mathrm{d}u}{1 + \sigma|u|^{n-1}u + (\sigma - 1)|u|^n} \tag{3.16}$$

$$\varphi_{\sigma,n}^+(w) = \int_0^w \frac{\mathrm{d}u}{1 - \sigma|u|^{n-1}u + (\sigma - 1)|u|^n} \tag{3.17}$$

$$\varphi_{\sigma,n}(w) = \varphi_{\sigma,n}^+(w) + \varphi_{\sigma,n}^-(w) \tag{3.18}$$

for any scalar $w \in (-1, 1)$. In this section and in the rest of the book the solution of the differential Equation (3.10) is denoted as $w(t)$, while the notation w without an argument is used for a given scalar. Note that for $w \geq 0$,

$$\varphi^{-}_{\sigma,n}(w) = \int_0^w \frac{du}{1+(2\sigma-1)u^n} \tag{3.19}$$

$$\varphi^{+}_{\sigma,n}(w) = \int_0^w \frac{du}{1-u^n} \tag{3.20}$$

$$\varphi_{\sigma,n}(w) = \int_0^w \frac{du}{1+(2\sigma-1)u^n} + \int_0^w \frac{du}{1-u^n} \tag{3.21}$$

and for $w \leq 0$,

$$\varphi^{-}_{\sigma,n}(w) = \int_0^w \frac{du}{1-(-u)^n} \tag{3.22}$$

$$\varphi^{+}_{\sigma,n}(w) = \int_0^w \frac{du}{1+(2\sigma-1)(-u)^n} \tag{3.23}$$

$$\varphi_{\sigma,n}(w) = \int_0^w \frac{du}{1-(-u)^n} + \int_0^w \frac{du}{1+(2\sigma-1)(-u)^n} \tag{3.24}$$

In the remainder of this section, some features of these functions and their respective inverse functions are presented. First, a check is made to ensure that the function $\varphi^{+}_{\sigma,n}$ is well defined, C^∞ and strictly increasing on the interval $[-1, 1)$. Consider the change of variable $v = u^n$. Then, for $w \geq 0.5$,

$$\begin{aligned} \varphi^{+}_{\sigma,n}(w) &= \varphi^{+}_{\sigma,n}(0.5) + \frac{1}{n}\int_{0.5^n}^{w^n} v^{1/n-1}\frac{dv}{1-v} \\ &\geq \varphi^{+}_{\sigma,n}(0.5) + \frac{0.5^{1/n-1}}{n}\int_{0.5^n}^{w^n} \frac{dv}{1-v} \end{aligned} \tag{3.25}$$

From Equation (3.25), the following limit property is drawn:

$$\lim_{w\to 1} \varphi^{+}_{\sigma,n}(w) = +\infty \tag{3.26}$$

All the attributes of the function $\varphi^{+}_{\sigma,n}$ stated above show that it is a bijection from $[-1, 1)$ to $[\varphi^{+}_{\sigma,n}(-1), +\infty)$. It is thus possible to define its inverse function

$$\psi^+_{\sigma,n} : [\varphi^+_{\sigma,n}(-1), +\infty)] \mapsto [-1, 1) \tag{3.27}$$

On the other hand, the function $\varphi^-_{\sigma,n}$ is well defined, C^∞ and strictly increasing on $(-1, 1]$. An analysis similar to that of Equation (3.25) shows that

$$\lim_{w \to -1} \varphi^-_{\sigma,n}(w) = -\infty \tag{3.28}$$

Therefore, the function $\varphi^-_{\sigma,n}$ is a bijection from $(-1, 1]$ to $(-\infty, \varphi^-_{\sigma,n}(1)]$. Then it is possible to define its inverse

$$\psi^-_{\sigma,n} : (-\infty, \ \varphi^-_{\sigma,n}(1)] \mapsto (-1, 1] \tag{3.29}$$

It can be checked that

$$\varphi^-_{\sigma,n}(w) = -\varphi^+_{\sigma,n}(-w) \qquad \text{and} \qquad \psi^-_{\sigma,n}(w) = -\psi^+_{\sigma,n}(-w) \tag{3.30}$$

Furthermore, the function $\varphi^+_{\sigma,n}$ is convex, while the function $\varphi^-_{\sigma,n}$ is concave.

For $u \in (0, 1)$,

$$\frac{1}{1 + (2\sigma - 1)u^n} < \frac{1}{1 - u^n} \tag{3.31}$$

Using this inequality in the integrals (3.19) and (3.20) gives

$$\varphi^+_{\sigma,n}(w) > \varphi^-_{\sigma,n}(w) \qquad \text{for } w \in (0, 1) \tag{3.32}$$

A similar analysis for the case of the integrals (3.22) and (3.23) shows that

$$\varphi^+_{\sigma,n}(w) > \varphi^-_{\sigma,n}(w) \qquad \text{for } w \in (-1, 0) \tag{3.33}$$

For the case of the function $\varphi_{\sigma,n}$, it can be checked that it is strictly increasing on the interval $(-1, 1)$ from $-\infty$ to $+\infty$, so that $\varphi_{\sigma,n}$ is a bijection from the interval $(-1, 1)$ to $\mathbb{R}$. Its inverse function is denoted as

$$\psi_{\sigma,n} : \mathbb{R} \mapsto (-1, 1) \tag{3.34}$$

Note that, from Equation (3.20), the restriction of the function $\varphi^{+}_{\sigma,n}$ to the interval $[0, 1)$ is independent of the parameter σ. For this reason, this restriction is denoted as $\varphi^{+}_{n} : [0, 1) \rightarrow \mathbb{R}^{+}$ with $\varphi^{+}_{n}(w) = \varphi^{+}_{\sigma,n}(w)$. Its inverse function is denoted as $\psi^{+}_{n} : \mathbb{R}^{+} \rightarrow [0, 1)$. Similarly, it can be seen from Equation (3.22) that the restriction of the function $\varphi^{-}_{\sigma,n}$ to the interval $(-1, 0]$ is independent of the parameter σ. For this reason, this restriction is denoted as $\varphi^{-}_{n} : (-1, 0] \rightarrow \mathbb{R}^{-}$ with $\varphi^{-}_{n}(w) = \varphi^{-}_{\sigma,n}(w)$. Its inverse function is denoted as $\psi^{-}_{n} : \mathbb{R}^{-} \rightarrow (-1, 0]$. These restrictions are defined by the equations

$$\varphi^{+}_{n}(w) = \varphi^{+}_{\sigma,n}(w) \qquad \text{for } w \geq 0 \tag{3.35}$$

$$\psi^{+}_{n}(\mu) = \psi^{+}_{\sigma,n}(\mu) \qquad \text{for } \mu \geq 0 \tag{3.36}$$

$$\varphi^{-}_{n}(w) = \varphi^{-}_{\sigma,n}(w) \qquad \text{for } w \leq 0 \tag{3.37}$$

$$\psi^{-}_{n}(\mu) = \psi^{-}_{\sigma,n}(\mu) \qquad \text{for } \mu \leq 0 \tag{3.38}$$

It can be remarked that, like the usual logarithm function, which is defined by an integral or its inverse, the exponential function, all the functions $\varphi^{-}_{\sigma,n}$, $\varphi^{+}_{\sigma,n}$, $\varphi_{\sigma,n}$, $\psi^{-}_{\sigma,n}$, $\psi^{+}_{\sigma,n}$, $\psi_{\sigma,n}$ are defined explicitly.

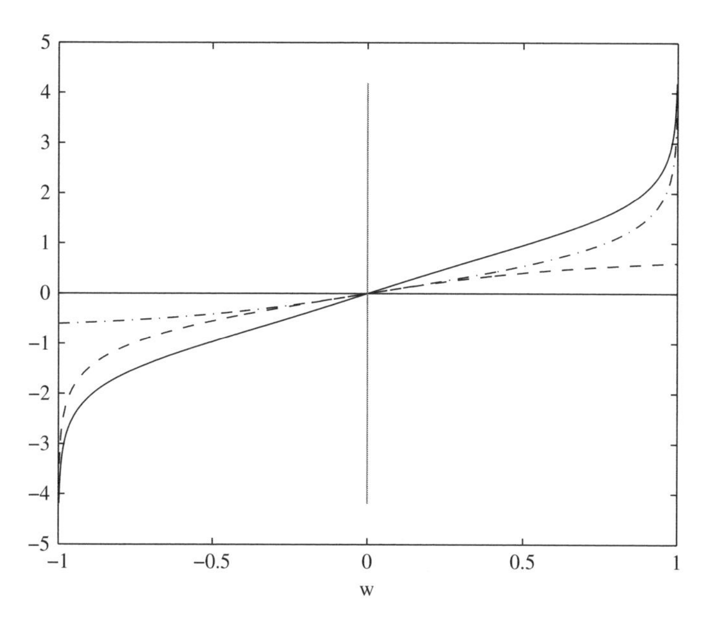

Figure 3.2 Functions $\varphi^{+}_{\sigma,n}(w)$ (dash-dot), $\varphi^{-}_{\sigma,n}(w)$ (dashed) and $\varphi_{\sigma,n}(w)$ (solid), with the values $\sigma = 2$ and $n = 2$.

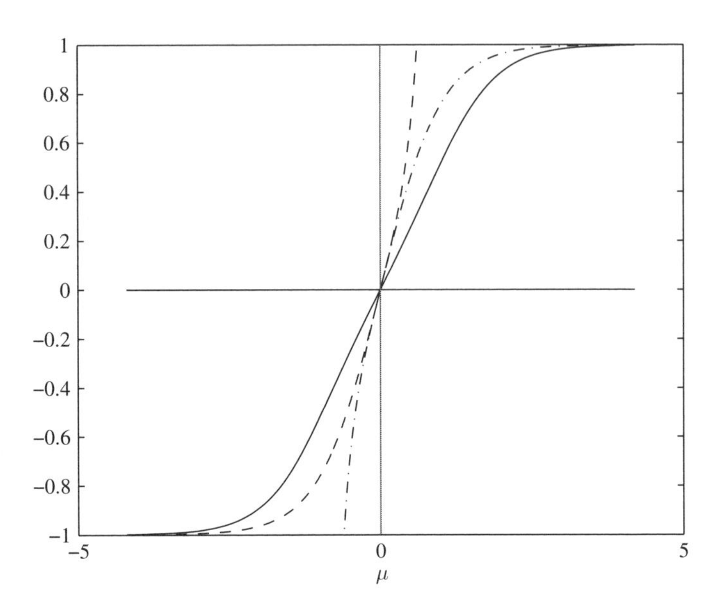

Figure 3.3 Functions $\psi^{+}_{\sigma,n}(\mu)$ (dash-dot), $\psi^{-}_{\sigma,n}(\mu)$ (dashed) and $\psi_{\sigma,n}(\mu)$ (solid), with the values $\sigma = 2$ and $n = 2$.

For example, to compute the function $\psi^{+}_{\sigma,n}$ numerically proceed as follows: given an argument $w \in [-1, 1)$ and the values of the parameters $\sigma \geq \frac{1}{2}$ and $n \geq 1$, the integrals (3.20) and (3.23) can be determined with the (in general very good) precision allowed by the computer. This means that for a given pair of parameters (σ, n) the function $\varphi^{+}_{\sigma,n}$ can be tabulated for a series of values $w_i \in [-1, 1)$, $i = 0, 1, 2, \ldots,$ as a series of pairs $\left(w_i, \varphi^{+}_{\sigma,n}(w_i)\right)$. Then the function $\psi^{+}_{\sigma,n}$ is tabulated by the pairs $\left(\varphi^{+}_{\sigma,n}(w_i), w_i\right)$, $i = 0, 1, 2, \ldots$. Intermediate values between two consecutive points i, $i+1$ are obtained by a linear interpolation. Examples of these functions are given in Figures 3.2 and 3.3.

3.5 CHARACTERIZATION OF THE ASYMPTOTIC BEHAVIOUR OF THE HYSTERETIC OUTPUT

In this section the normalized Bouc–Wen model (3.9)–(3.10) is considered to be excited by a wave periodic signal x. It is shown that the hysteretic output $\Phi_{\mathrm{BW}}(x)$ converges asymptotically to a periodic function. The main result of this section is given in the following theorem.

Theorem 3. *Consider the system (3.9)–(3.10) with an initial condition $w(0)$, where the input signal x is wave T-periodic (see Section 3.2.1). Define the functions ω_m and ϕ_m for any nonnegative integer m as follows:*

$$\omega_m(\tau) = w(t_m + \tau) \qquad \text{for } \tau \in [0, T] \tag{3.39}$$

$$\phi_m(\tau) = \kappa_x x(\tau) + \kappa_w \omega_m(\tau) \text{ for } \tau \in [0, T] \tag{3.40}$$

These give the following:

(*a*) *The sequences of functions $\{\phi_m\}_{m\geq 0}$ and $\{\omega_m\}_{m\geq 0}$ converge uniformly on the interval $[0, T]$ to the continuous functions $\bar{\Phi}_{BW}$ and $\bar{w}$ defined in the form*

$$\bar{\Phi}_{BW}(\tau) = \kappa_x x(\tau) + \kappa_w \bar{w}(\tau) \qquad \text{for } \tau \in [0, T] \tag{3.41}$$

$$\bar{w}(\tau) = \psi^{+}_{\sigma,n}\left(\varphi^{+}_{\sigma,n}\left[-\psi_{\sigma,n}\left(\rho\left(X_{\max} - X_{\min}\right)\right)\right] + \rho\left(x(\tau) - X_{\min}\right)\right) \quad \text{for } \tau \in [0, T^{+}] \tag{3.42}$$

$$\bar{w}(\tau) = -\psi^{+}_{\sigma,n}\left(\varphi^{+}_{\sigma,n}\left[-\psi_{\sigma,n}\left(\rho\left(X_{\max} - X_{\min}\right)\right)\right] - \rho\left(x(\tau) - X_{\max}\right)\right) \quad \text{for } \tau \in [T^{+}, T] \tag{3.43}$$

(*b*) *For all $\tau \in [0, T]$,*

$$-1 < -\psi_{\sigma,n}\left(\rho\left(X_{\max} - X_{\min}\right)\right) \leq \bar{w}(\tau) \leq \psi_{\sigma,n}\left(\rho\left(X_{\max} - X_{\min}\right)\right) < 1 \tag{3.44}$$

the lower and upper bounds of $\bar{w}(\tau)$ being attained at $\tau = 0$ and $\tau = T^{+}$ respectively.

The lengthy proof of this theorem is given in Sections 3.5.1 and 3.5.2. Section 3.5.1 presents some technical lemmas that are used in the proof of Theorem 3. Section 3.5.2 presents the proof of Theorem 3 using the lemmas of Section 3.5.1.

Some comments are now given regarding Theorem 3. Define the time function $\bar{\phi}_{BW}$ as $\bar{\phi}_{BW}(t) = \bar{\Phi}_{BW}(\tau)$ where the time $t \in [0, +\infty)$ is written as $t = mT + \tau$ for all integers $m = 0, 1, 2, \ldots$ and all real numbers $0 \leq \tau < T$. Loosely speaking, Theorem 3 says that the time function hysteretic output $\Phi_{BW}(x)(t)$ of the Bouc–Wen model approaches asymptotically the T-periodic function $\bar{\phi}_{BW}(t)$. The limit cycle is the graph $\left(x(\tau), \bar{\Phi}_{BW}(\tau)\right)$ parameterized by the variable $0 \le \tau \le T$. Equations (3.41) and (3.42) correspond to the so-called loading, that is to an increasing input $x(\tau)$ with $0 \le \tau \le T^+$. Equations (3.41) and (3.43) correspond to the unloading, that is to a decreasing input $x(\tau)$ with $T^+ \le \tau \le T$.

The equations of the limit cycle are given in terms of the variable τ. However, what is of interests is the relation $\bar{\Phi}_{\mathrm{BW}}(x)$ where the variable τ is eliminated. To see this elimination process, consider the following example.

Example 2. *Consider the curve $(x(\tau), y(\tau))$ which is parameterized by the variable τ such that*

$$x(\tau) = \frac{\tau}{2} + 3 \tag{3.45}$$

$$y(\tau) = \tau^3 \tag{3.46}$$

To find the equation of the curve in the plane (x, y), proceed as follows. Equation (3.45) gives $\tau = 2\,(x-3)$ while Equation (3.46) gives $\tau = \sqrt[3]{y}$. Equating both quantities gives $\sqrt[3]{y} = 2\,(x-3)$, which is the desired equation of the curve in the (x, y) plane.

Let us first consider Equations (3.41) and (3.42) which describe the loading. In the interval $[0, T^+]$, the input displacement is an increasing function of the variable τ; that is $x = f(\tau)$ for some function f. Since the function f is increasing, it is invertible so that $\tau = f^{-1}(x)$, where f^{-1} is the inverse function of f. Now, consider the function g defined as

$$g(u) = \psi^+_{\sigma,n}\left(\varphi^+_{\sigma,n}\left[-\psi_{\sigma,n}\left(\rho\left(X_{\max} - X_{\min}\right)\right)\right] + \rho\left(u - X_{\min}\right)\right) \tag{3.47}$$

for $u \in [X_{\min}, X_{\max}]$. The term $\varphi^+_{\sigma,n}\left[-\psi_{\sigma,n}\left(\rho\left(X_{\max} - X_{\min}\right)\right)\right]$ is independent of u so that the function $g(u)$ is increasing with u. This is due to the fact that the function $\psi^+_{\sigma,n}(\cdot)$ is increasing with its argument (Section 3.4). The fact that the function g is increasing implies that it is invertible. Now, combining Equations (3.42) and (3.47), it follows that

$$\bar{w}(\tau) = g\left(f(\tau)\right) = \left(g \circ f\right)(\tau) \tag{3.48}$$

Since both f and g are invertible, from Equation (3.48) it is found that $\tau = \left(g \circ f\right)^{-1}(\bar{w})$. The equation of the loading part of the limit cycle is obtained by equating both expressions of τ so that $(g \circ f)^{-1}(\bar{w}) = f^{-1}(x)$, which leads to $\bar{w} = g(x)$. Similarly, it can be shown that $\bar{\Phi}_{\mathrm{BW}} = \kappa_x x + \kappa_w g(x)$. Thus, the loading equations (3.41) and (3.42) can be rewritten as

$$\bar{\Phi}^{\mathrm{l}}_{\mathrm{BW}}(x) = \kappa_x x + \kappa_w \bar{w}^{\mathrm{l}}(x) \tag{3.49}$$

$$\bar{w}^{\mathrm{l}}(x) = \psi^{+}_{\sigma,n}\left(\varphi^{+}_{\sigma,n}\left[-\psi_{\sigma,n}\left(\rho\left(X_{\max} - X_{\min}\right)\right)\right] + \rho\left(x - X_{\min}\right)\right) \tag{3.50}$$

where the superscript 'l' refers to loading. In Equations (3.49) and (3.50), $\bar{\Phi}^{\mathrm{l}}_{\mathrm{BW}}(x)$ and $\bar{w}^{\mathrm{l}}(x)$ are functions of the input signal x.

Similarly, unloading is described by the equations

$$\bar{\Phi}^{\mathrm{u}}_{\mathrm{BW}}(x) = \kappa_x x + \kappa_w \bar{w}^{\mathrm{u}}(x) \tag{3.51}$$

$$\bar{w}^{\mathrm{u}}(x) = -\psi^{+}_{\sigma,n}\left(\varphi^{+}_{\sigma,n}\left[-\psi_{\sigma,n}\left(\rho\left(X_{\max} - X_{\min}\right)\right)\right] - \rho\left(x - X_{\max}\right)\right) \tag{3.52}$$

where the superscript 'u' refers to unloading.

Consequently, seen as a function of $x \in [X_{\min}, X_{\max}]$, loading and unloading are described respectively by the functions

$$\bar{\Phi}^{\mathrm{l}}_{\mathrm{BW}}(x) = \kappa_x x + \kappa_w \psi^{+}_{\sigma,n}\left(\varphi^{+}_{\sigma,n}\left[-\psi_{\sigma,n}\left(\rho\left(X_{\max} - X_{\min}\right)\right)\right] + \rho\left(x - X_{\min}\right)\right) \tag{3.53}$$

$$\bar{\Phi}^{\mathrm{u}}_{\mathrm{BW}}(x) = \kappa_x x - \kappa_w \psi^{+}_{\sigma,n}\left(\varphi^{+}_{\sigma,n}\left[-\psi_{\sigma,n}\left(\rho\left(X_{\max} - X_{\min}\right)\right)\right] - \rho\left(x - X_{\max}\right)\right) \tag{3.54}$$

The functions $\bar{\Phi}^{\mathrm{l}}_{\mathrm{BW}}$ and $\bar{\Phi}^{\mathrm{u}}_{\mathrm{BW}}$ describe completely the limit cycle, which are C^{∞} on the interval $[X_{\min}, X_{\max}]$. It can be seen from Equations (3.53) and (3.54) that the functions $\bar{\Phi}^{\mathrm{l}}_{\mathrm{BW}}$ and $\bar{\Phi}^{\mathrm{u}}_{\mathrm{BW}}$ are independent of the period T and the initial state $w(0)$. The fact that the limit cycle is independent of the frequency of the input signal is called the 'rate-independent' property and is an inherent property of the physical hysteretic systems (at least for low frequencies) [126, page xiv]. A by-product of Theorem 3 is the fact that the Bouc–Wen model is rate independent for all frequencies.

3.5.1 Technical Lemmas

Section 3.4 has presented the functions $\varphi^{-}_{\sigma,n}$, $\varphi^{+}_{\sigma,n}$, $\varphi_{\sigma,n}$ and it was seen that they are all well defined on the interval $(-1, 1)$. These functions will be shown to be extremely useful for integrating Equation (3.10)and describing the forced limit cycles of the hysteretic Bouc–Wen model. However, to be used, the argument of these functions should belong to the interval $(-1, 1)$. The following lemma is instrumental in guaranteeing such a condition.

Lemma 3. *For any initial condition* $w(0) \in \Omega$, *a finite time instant* $t_{|w|<1}$ *exists such that for every* $t \geq t_{|w|<1}$ *then* $|w(t)| < 1$.

Proof. The proof is done in two steps:

1. First the existence of a time instant $t_{|w|\leq 1}$ such that is established $|w(t)| \leq 1$ for all $t \geq t_{|w|\leq 1}$.
2. Then it is shown that some integer m_0 with $t_{m_0} \geq t_{|w|\leq 1}$ exists for which $|w_{m_0}| < 1$, and that $t_{|w|<1} = t_{m_0}$ can be taken.

Proof of step 1. Assume that $w(t) > 1$ for all $t \geq 0$ (a similar analysis can be done in the case $w(t) < -1$). From Equation (3.8), it follows that

$$w^2(t) = 2\frac{V(t)}{z_0^2} \qquad \text{where } V(t) = \frac{z^2(t)}{2}$$

On the other hand, it has been demonstrated in Section 2.2.3 that the time function $V(t)$ is nonincreasing for a class I Bouc–Wen model (which is the present case) whenever $z^2(t) \geq z_0^2$. This fact implies that the function $w(t)^2$ is nonincreasing. Since it is bounded from below as $w(t) > 1$, it goes to a limit $\ell \geq 1$.

Assume now that this limit is $\ell > 1$. Then on the intervals I_m^+, $m \geq 0$ (see Figure 3.1), $\dot{x} \geq 0$, so that Equation (3.10) reduces to Equation (3.12). Integrating Equation (3.12) from t_m to t_m^+ gives

$$\int_{w_m}^{w_m^+} \frac{du}{u^n - 1} = -\delta \tag{3.55}$$

where

$$\delta = \rho\,(X_{\max} - X_{\min}) > 0 \tag{3.56}$$

Since $\lim_{t\to\infty} w(t) = \lim_{m\to\infty} w_m = \lim_{m\to\infty} w_m^+ = \ell$, then taking the limit in the left-hand side of Equation (3.55) gives $\delta = 0$, which contradicts Equation (3.56). This means that it is not possible to have $\ell > 1$. Since $\ell \geq 1$, then $\ell = 1$.

Now, for the limit $\ell = 1$, take $\varepsilon > 0$. Then some time τ_ε exists such that for all $t \geq \tau_\varepsilon$

$$1 - \varepsilon \leq w(t)^n \leq 1 + \varepsilon \tag{3.57}$$

An interval I_m^- (see Figure 3.1) can be considered such that $t_m^+ \geq \tau_\varepsilon$ and Equation (3.14) used as $w(t) \geq 0$ and $\dot{x} \leq 0$. Integrating Equation (3.14) between t_m^+ and t_{m+1} and using (3.57) gives

$$w_{m+1} - w_m^+ \leq \rho \left(X_{\min} - X_{\max}\right) [2\sigma - \varepsilon(2\sigma - 1)] < 0 \tag{3.58}$$

Since the difference $w_{m+1} - w_m^+$ goes to 0 as m go to ∞, for any ε sufficiently small it will be a value of m for which the expression (3.58) will be violated, thus giving a contradiction.

Therefore it has been proved that it is impossible to have $|w(t)| > 1$ for all $t \geq 0$. This means that for any initial condition $w(0) \in \Omega$, some finite time $t_{|w|\leq 1}$ for which we have $\left|w\left(t_{|w|\leq 1}\right)\right| \leq 1$. From Table 3.1 it follows that for all $t \geq t_{|w|\leq 1}$ we have $|w(t)| \leq 1$.

Proof of step 2. Two cases are considered:

1. For all $t \geq t_{|w|\leq 1}$, then $|w(t)| < 1$.
2. A time instant $t_{|w|=1} \geq t_{|w|\leq 1}$ exists such that $\left|w\left(t_{|w|=1}\right)\right| = 1$.

In the first case, Lemma 3 follows by taking $t_{|w|<1} = t_{|w|\leq 1}$. In the second case, take for example $w\left(t_{|w|=1}\right) = 1$ (the same analysis holds for $w\left(t_{|w|=1}\right) = -1$). Then two subcases are to be discussed:

(a) $t_{|w|=1} \in [t_p, t_p^+)$ for some integer p.
(b) $t_{|w|=1} \in [t_p^+, t_{p+1})$ for some integer p.

In the subcase (a), $w(t) = 1$ for all $t \in [t_{|w|=1}, t_p^+]$ as the point defined by $w(t) = 1$ and $\dot{w}(t) = 0$ is an equilibrium point for Equation (3.10). Integrating Equation (3.10) on I_p^+ results in

$$\varphi_{\sigma,n}^-\left(w_{p+1}\right) = \varphi_{\sigma,n}^-(1) - \delta \tag{3.59}$$

From Equation (3.59) it follows that the value $\varphi_{\sigma,n}^-\left(w_{p+1}\right)$ is finite; this means that $-1 < w_{p+1} \leq 1$. Furthermore, $\varphi_{\sigma,n}^-\left(w_{p+1}\right) < \varphi_{\sigma,n}^-(1)$

by Equation (3.59). Thus it is concluded that $|w_{p+1}| < 1$. The same inequality can be obtained for the subcase (b).

It has therefore been proved that an integer $m_0 = p+1$ exists such that $|w_{m_0}| < 1$. Assume now that for some integer $m \geq m_0$ then $|w_m| < 1$. The objective of the following analysis is to demonstrate that for all $t \in [t_m, t_{m+1}]$ then $|w(t)| < 1$. To this end, note that for $t \in I_m^+$ then, from Equation (3.10),

$$\varphi_{\sigma,n}^{+}(w(t)) = \varphi_{\sigma,n}^{+}(w_m) + \rho\,(x(t) - X_{\min}) \tag{3.60}$$

Equation (3.60) shows that the value $\varphi_{\sigma,n}^{+}(w(t))$ is finite, which implies that

$$-1 < w_m \leq w(t) \leq w_m^+ < 1 \qquad \text{for} \quad t \in I_m^+$$

Now, on the interval I_m^-, from equation (3.10),

$$\varphi_{\sigma,n}^{-}(w(t)) = \varphi_{\sigma,n}^{-}(w_m^+) + \rho\,(x(t) - X_{\max}) \tag{3.61}$$

Again, Equation (3.61) shows that the value $\varphi_{\sigma,n}^{-}(w(t))$ is finite, which implies that

$$-1 < w_{m+1} \leq w(t) \leq w_m^+ < 1 \qquad \text{for} \quad t \in I_m^-$$

Thus it has been proved that, for all $t \in [t_m, t_{m+1}]$, then $|w(t)| < 1$. By taking $t_{|w|<1} = t_{m_0}$, Lemma 3 is finally proved.

Since we are interested only in asymptotic behaviour, it is enough to ensure that $w(t)$ will belong to the interval $(-1, 1)$ for t sufficiently large. Lemma 3 shows that for all $t \geq t_{|w|<1}$ the argument $w(t)$ of the functions $\varphi_{\sigma,n}^{-}$, $\varphi_{\sigma,n}^{+}$, $\varphi_{\sigma,n}$ belongs to their domain of definition. This will allow the analysis of the asymptotic behaviour of $w(t)$ in the remainder of this section.

Lemma 4. *For any initial condition $w(0) \in \Omega$, let $t_{|w|<1}$ be the time instant of Lemma 3. Then, some time instant $t_{w=0} \geq t_{|w|<1}$ exists for which $w(t_{w=0}) = 0$.*

Proof. Assume that $w(t) > 0$ for all $t \geq t_{|w|<1}$ (the analysis is similar for $w(t) < 0$ for all $t \geq t_{|w|<1}$) and let m_0 be an integer such that $t_{m_0} \geq t_{|w|<1}$. Then, for all integer m such that $m \geq m_0$ the following holds:

$$\varphi_{\sigma,n}^{+}(w_m^+) = \varphi_{\sigma,n}^{+}(w_m) + \delta \tag{3.62}$$

$$\varphi_{\sigma,n}^{-}(w_{m+1}) = \varphi_{\sigma,n}^{-}(w_m^{+}) - \delta \tag{3.63}$$

The real sequence $\{w_m\}_{m\geq m_0}$ is bounded, so that a convergent subsequence $\{w_{m_p}\}_{p\geq 0}$ can be extracted from it. The real sequence $\{w_{m_p}^{+}\}_{p\geq 0}$ is bounded, so that a convergent subsequence $\{w_{m_{p_k}}^{+}\}_{k\geq 0}$ can be extracted. Denote

$$\lim_{k\to\infty} w_{m_{p_k}} = w_a \qquad \text{and} \qquad \lim_{k\to\infty} w_{m_{p_k}}^{+} = w_b$$

Then, by the continuity of the functions $\varphi_{\sigma,n}^{+}$ and $\varphi_{\sigma,n}^{-}$,

$$\varphi_{\sigma,n}^{+}(w_b) = \varphi_{\sigma,n}^{+}(w_a) + \delta \tag{3.64}$$

$$\varphi_{\sigma,n}^{-}(w_a) = \varphi_{\sigma,n}^{-}(w_b) - \delta \tag{3.65}$$

From Equation (3.65) it follows that $0 \leq w_a < w_b \leq 1$, so that $\varphi_{\sigma,n}^{+}(w_a)$ is finite. This also implies by Equation (3.64) that $\varphi_{\sigma,n}^{+}(w_b)$ is finite, so that $w_b < 1$.

Note that it is not possible to have $w_a = 0$ as this would mean from Equations (3.64) and (3.65) that $\varphi_{\sigma,n}^{+}(w_b) = \varphi_{\sigma,n}^{-}(w_b)$. This cannot happen as $w_b > 0$ and $\sigma > 0$. This means that $0 < w_a < w_b < 1$. Now, from Equations (3.64) and (3.65), from the convexity of $\varphi_{\sigma,n}^{+}$ and the concavity of $\varphi_{\sigma,n}^{-}$,

$$\begin{aligned} \frac{1}{1-w_a^n} = \left[\frac{d\varphi_{\sigma,n}^{+}(w)}{dw}\right]_{w=w_a} &< \frac{\varphi_{\sigma,n}^{+}(w_b) - \varphi_{\sigma,n}^{+}(w_a)}{w_b - w_a} \\ &= \frac{\delta}{w_b - w_a} \end{aligned} \tag{3.66}$$

$$\begin{aligned} \frac{\varphi_{\sigma,n}^{-}(w_b) - \varphi_{\sigma,n}^{-}(w_a)}{w_b - w_a} = \frac{\delta}{w_b - w_a} &\leq \left[\frac{d\varphi_{\sigma,n}^{-}(w)}{dw}\right]_{w=w_a} \\ &= \frac{1}{1+(2\sigma-1)w_a^n} \end{aligned} \tag{3.67}$$

From Equations (3.66) and (3.67),

$$w_a + \delta\left(1 + (2\sigma - 1)w_a^n\right) \leq w_b < w_a + \delta\left(1 - w_a^n\right) \tag{3.68}$$

This gives a contradiction as $w_a > 0$ and $\sigma \geq 1/2$, which means that for a class I Bouc–Wen model there cannot be $w(t) > 0$ for all $t \geq t_{|w|<1}$.

Lemma 5. *For any initial condition $w(0) \in \Omega$, let $t_{|w|<1}$ be the time instant of Lemma 3. Let m_0 be an integer such that $t_{m_0} \geq t_{|w|<1}$. Then for all integers $m \geq m_0$, the following holds: in any time interval $[t_m, t_m^+)$ or $[t_m^+, t_{m+1})$, the equation $w(t) = 0$ has at most one solution.*

Proof. Assume that this is not the case, so that $w(\tau_1) = w(\tau_2) = 0$, where $t_m \leq \tau_1 < \tau_2 < t_m^+$. Then some $\tau_1 < \tau_3 < \tau_2$ would exist such that $\dot{w}(\tau_3) = 0$. Since $\dot{w}(t)$ is given by Equation (3.10) and since $t_m < \tau_3 < t_m^+$, then $\dot{x}(\tau_3) > 0$ so that

$$1 - \sigma|w(\tau_3)|^{n-1}w(\tau_3) + (\sigma - 1)|w(\tau_3)|^n = 0 \tag{3.69}$$

It can be checked that no solution $|w(\tau_3)| < 1$ of Equation (3.69) exists. The same can be said for the interval $[t_m^+, t_{m+1})$.

Lemma 6. *For any initial condition $w(0) \in \Omega$, let $t_{|w|<1}$ be the time instant of Lemma 3. Let m_1 be an integer such that $t_{m_1} \geq t_{|w|<1}$. If $-\delta < \varphi_{\sigma,n}^-(w_{m_1}) < 0$ then $-\delta < \varphi_{\sigma,n}^-(w_m) < 0$ and $0 < \varphi_{\sigma,n}^+(w_m^+) < \delta$ for each integer $m \geq m_1$. Furthermore, the equation $w(t) = 0$ has exactly one solution in each interval (t_m, t_m^+) and (t_m^+, t_{m+1}).*

Proof. Assume that for some $m \geq m_1$ then $-\delta < \varphi_{\sigma,n}^-(w_m) < 0$. For $t \in I_m^+$, the following is found by integrating Equation (3.10):

$$\varphi_{\sigma,n}^+(w(t)) - \varphi_{\sigma,n}^+(w_m) = \rho\,(x(t) - X_{\min}) \tag{3.70}$$

From Equation (3.70) and that fact that the function $x(t)$ is strictly increasing on the interval I_m^+, it follows that the time function $w(t)$ is strictly increasing from $w_m < 0$ to w_m^+, and

$$0 < \varphi_{\sigma,n}^-(w_m) + \delta < \varphi_{\sigma,n}^+(w_m) + \delta = \varphi_{\sigma,n}^+(w_m^+) < \delta \tag{3.71}$$

Equation (3.71) implies that $w_m^+ > 0$ and that a unique solution exists to the equation $w(t) = 0$ in the interval (t_m, t_m^+) as $w(t)$ is continuous. Now, integrating Equation (3.10) on the interval I_m^- gives

$$\varphi_{\sigma,n}^-(w(t)) - \varphi_{\sigma,n}^-(w_m^+) = \rho\,(x(t) - X_{\max}) \tag{3.72}$$

Here the time function $w(t)$ is strictly decreasing from $w_m^+ > 0$ to $w_{m+1} < 0$ such that

$$-\delta < \varphi_{\sigma,n}^-(w_{m+1}) = \varphi_{\sigma,n}^-(w_m^+) - \delta < \varphi_{\sigma,n}^+(w_m^+) - \delta < 0$$

This means that $w_{m+1} < 0$ and that a unique solution exists to the equation $w(t) = 0$ in the interval (t_m^+, t_{m+1}), which proves Lemma 6.

Lemma 7. *For any initial condition $w(0) \in \Omega$ let $t_{|w|<1}$ be the time instant of Lemma 3. Some finite integer $m_1 \geq 0$ exists such that for all integers $m \geq m_1$ the following holds:*

(*a*) *The equation $w(t) = 0$ has exactly one solution in each interval (t_m, t_m^+). Denote it by $t_{m,w=0}^+$.*
(*b*) *The equation $w(t) = 0$ has exactly one solution in each interval (t_m^+, t_{m+1}). Denote it by $t_{m,w=0}^-$.*
(*c*) *Define*

$$-1 < \lambda_{\min} = \psi_{\sigma,n}^-(-\delta) < 0 \quad \text{and} \quad 0 < \lambda_{\max} = \psi_{\sigma,n}^+(\delta) < 1$$

Then

$$\begin{aligned} \lambda_{\min} < w(t) < 0 \qquad & \text{for } t \in (t_m, t_{m,w=0}^+) \\ 0 < w(t) < \lambda_{\max} \qquad & \text{for } t \in (t_m^+, t_{m,w=0}^-) \\ 0 < w(t) < \lambda_{\max} \qquad & \text{for } t \in (t_{m,w=0}^+, t_m^+) \\ \lambda_{\min} < w(t) < 0 \qquad & \text{for } t \in (t_{m,w=0}^-, t_{m+1}) \end{aligned}$$

Proof. It is known by Lemma 4 that some time instant $t_{w=0} \geq t_{|w|<1}$ exists for which $w(t_{w=0}) = 0$. There are two possibilities:

1. $t_{w=0} \in [t_m, t_m^+)$ for some $m \geq m_0$, where $t_{m_0} \geq t_{|w|<1}$.
2. $t_{w=0} \in [t_m^+, t_{m+1})$.

Consider the first case. Then

$$\dot{w}(t) = \rho\left(1 - \sigma|w(t)|^{n-1}w(t) + (\sigma - 1)|w(t)|^n\right)\dot{x}(t) \qquad \text{for } t \in I_m^+$$

For sufficiently small ε, then $\dot{w}(t) > 0$ for $t \in (t_{w=0}, t_{w=0} + \varepsilon)$, so that by Lemma 5 and by the continuity of $w(t)$

$$w(t) \leq 0 \qquad \text{for } t_m \leq t \leq t_{w=0}$$

and

$$w(t) > 0 \qquad \text{for } t_{w=0} < t \leq t_m^+$$

This means in particular that $w_m^+ > 0$ and $w_m \leq 0$. Integrating Equation (3.10) in the time interval $[t_m, t_{w=0}]$ and taking into account the fact that $x(t_{w=0}) < X_{\max}$ gives

$$-\delta < \varphi_{\sigma,n}^+(w_m) = \rho\,(X_{\min} - x\,(t_{w=0})) \leq 0 \tag{3.73}$$

Integrating Equation (3.10) in the time interval $[t_{w=0}, t_m^+]$ gives

$$\varphi_{\sigma,n}^+(w_m^+) = \rho\,(X_{\max} - x\,(t_{w=0})) \geq 0 \tag{3.74}$$

Since $w_m^+ > 0$ and $x\,(t_{w=0}) \geq X_{\min}$, then $0 < \varphi_{\sigma,n}^+(w_m^+) \leq \delta$ from Equation (3.74). Now, integrating Equation (3.10) in the time interval $[t_m^+, t_{m+1}]$ gives

$$\varphi_{\sigma,n}^-(w_{m+1}) = \varphi_{\sigma,n}^-(w_m^+) - \delta \tag{3.75}$$

Since $\varphi_{\sigma,n}^-(w_m^+) < \varphi_{\sigma,n}^+(w_m^+) \leq \delta$, then $\varphi_{\sigma,n}^-(w_{m+1}) < 0$ by Equation (3.75). On the other hand, since $w_m^+ > 0$, then $\varphi_{\sigma,n}^-(w_m^+) > 0$ so that, by Equation (3.75), $-\delta < \varphi_{\sigma,n}^-(w_{m+1}) < 0$ is obtained. This gives the conditions of Lemma 6 by taking $m_1 = m_0 + 1$, from which Lemma 7 follows.

Case 2 is treated similarly.

3.5.2 Analytic Description of the Forced Limit Cycles for the Bouc–Wen Model

This section uses the technical lemmas of Section 3.5.1 to demonstrate Theorem 3.

Proof. Let m_1 be an integer as in Lemma 7. Fix some $\tau \in [0, T^+]$ and take $m \geq m_1$. Then, by integrating Equation (3.10) in the appropriate intervals, the following from Lemma 7 is obtained:

$$\varphi_{\sigma,n}^+(w_m^+) - \varphi_{\sigma,n}^+\,(w(t_m + \tau)) = \rho\,(X_{\max} - x(\tau)) \tag{3.76}$$

$$\varphi_{\sigma,n}^-(w_{m+1}) - \varphi_{\sigma,n}^-\,(w_m^+) = -\delta \tag{3.77}$$

$$\varphi_{\sigma,n}^+\,(w(t_{m+1} + \tau)) - \varphi_{\sigma,n}^+\,(w_{m+1}) = \rho\,(x(\tau) - X_{\min}) \tag{3.78}$$

From Equation (3.76), it follows that

$$w_m^+ = \psi_{\sigma,n}^+ \left[\varphi_{\sigma,n}^+ (w(t_m + \tau)) + \rho\, (X_{\max} - x(\tau))\right] \triangleq g_1\, (w(t_m + \tau)) \quad (3.79)$$

Since the variable τ has been fixed, g_1 is a function of the variable $w(t_m + \tau)$ (this variable changes with m). The function g_1 is increasing as the functions $\psi_{\sigma,n}^+$ and $\varphi_{\sigma,n}^+$ are also increasing.

From Equation (3.77), it follows that

$$w_{m+1} = \psi_{\sigma,n}^- \left[\varphi_{\sigma,n}^- (w_m^+) - \delta\right] \triangleq g_2(w_m^+) \quad (3.80)$$

The function g_2 is increasing as the functions $\psi_{\sigma,n}^-$ and $\varphi_{\sigma,n}^-$ are also increasing.

From Equation (3.78), it follows that

$$w\,(t_{m+1} + \tau) = \psi_{\sigma,n}^+ \left[\varphi_{\sigma,n}^+ (w_{m+1}) + \rho\, (x(\tau) - X_{\min})\right] \triangleq g_3(w_{m+1}) \quad (3.81)$$

Since the variable τ has been fixed, g_3 is a function of the variable w_{m+1} (this variable changes with m). The function g_3 is increasing as the functions $\psi_{\sigma,n}^+$ and $\varphi_{\sigma,n}^+$ are also increasing.

From Equations (3.79), (3.80) and (3.81), it follows that

$$w\,(t_{m+1} + \tau) = (g_3 \circ g_2 \circ g_1)\,[w(t_m + \tau)] \triangleq f\,(w(t_m + \tau)) \quad (3.82)$$

The function f is increasing as the functions g_1, g_2 and g_3 are also increasing.

Let us now consider the two cases:

1. $w(t_{m_1+1} + \tau) \geq w(t_{m_1} + \tau)$.
2. $w(t_{m_1+1} + \tau) \leq w(t_{m_1} + \tau)$.

First case 1 is treated. Since the function f is increasing, then

$$f\left(w(t_{m_1+1} + \tau)\right) \geq f\left(w(t_{m_1} + \tau)\right)$$

This implies that

$$w(t_{m_1+2} + \tau) \geq w(t_{m_1+1} + \tau)$$

Thus, by induction, it follows that, for all integers $m \geq m_1$,

$$w(t_{m+1} + \tau) \geq w(t_m + \tau)$$

Thus the real sequence $\{w(t_m+\tau)\}_{m\geq m_1}$ is monotonic. The same conclusion (monotonicity of the sequence $\{w(t_m+\tau)\}_{m\geq m_1}$) is obtained for case 2. This implies that, in all cases, the real sequence $\{w(t_m+\tau)\}_{m\geq m_1}$ is monotonic. Since this sequence is bounded, it goes to a limit $\bar{w}(\tau)$ which, by Lemma 7, verifies

$$-1 < \lambda_{\min} \leq \bar{w}(\tau) \leq \lambda_{\max} < 1$$

Taking $\tau = 0$ in Equations (3.76) to (3.78) gives

$$\varphi^{+}_{\sigma,n}\left(\bar{w}(T^{+})\right) - \varphi^{+}_{\sigma,n}\left(\bar{w}(0)\right) = \delta \tag{3.83}$$

$$\varphi^{-}_{\sigma,n}\left(\bar{w}(T^{+})\right) - \varphi^{-}_{\sigma,n}\left(\bar{w}(0)\right) = \delta \tag{3.84}$$

From Lemma 7, it follows that

$$\bar{w}(0) \leq 0 \leq \bar{w}(T^{+}) < 1$$

From Equations (3.83) and (3.84) it follows that

$$-1 < \bar{w}(0) < \bar{w}(T^{+}) < 1$$

Denoting

$$g(\cdot) = \varphi^{+}_{\sigma,n}(\cdot) - \varphi^{-}_{\sigma,n}(\cdot)$$

gives $g\left(\bar{w}(T^{+})\right) = g\left(\bar{w}(0)\right)$ from Equations (3.83) and (3.84). It can be shown easily that $g(\cdot)$ is strictly increasing on the interval (0,1), strictly decreasing on the interval $(-1, 0)$ and is even on the interval $(-1, 1)$, so that necessarily $\bar{w}(0) = -\bar{w}(T^{+})$. Equations (3.83) and (3.84) then reduce to the single relation

$$\varphi^{+}_{\sigma,n}\left(\bar{w}(T^{+})\right) + \varphi^{-}_{\sigma,n}\left(\bar{w}(T^{+})\right) = \varphi_{\sigma,n}\left(\bar{w}(T^{+})\right) = \delta \tag{3.85}$$

Then from Equation (3.85),

$$\bar{w}(T^{+}) = \psi_{\sigma,n}(\delta) \qquad \text{and} \qquad \bar{w}(0) = -\psi_{\sigma,n}(\delta)$$

Now, fixing $\tau \in [T^{+}, T]$ gives the following equations:

$$\varphi^{-}_{\sigma,n}(w_{m+1}) - \varphi^{-}_{\sigma,n}\left(w(t_m+\tau)\right) = \rho\left(X_{\min} - x(\tau)\right) \tag{3.86}$$

$$\varphi_{\sigma,n}^{+}(w_{m+1}^{+}) - \varphi_{\sigma,n}^{+}(w_{m+1}) = \delta \tag{3.87}$$

$$\varphi_{\sigma,n}^{-}(w(t_{m+1}+\tau)) - \varphi_{\sigma,n}^{-}(w_{m+1}^{+}) = \rho\,(x(\tau) - X_{\max}) \tag{3.88}$$

Similar to the analysis above, it is conclude that the real sequence $\{w(t_m+\tau)\}_{m\geq m_1}$ goes to a limit $\bar{w}(\tau)$ which, again by Lemma 7, verifies

$$-1 < \lambda_{\min} \leq \bar{w}(\tau) \leq \lambda_{\max} < 1$$

Equations (3.42) and (3.43) of Theorem 3 follow by taking $m \to +\infty$ in Equations (3.76) to (3.78) and Equations (3.86) to (3.88). It can be checked that $\bar{w}(0) = \bar{w}(T)$ and that the expressions of $\bar{w}(T^{+})$ given by Equations (3.42) and (3.43) are equal so that the function $\bar{w}(\tau)$ is continuous.

The analysis above has shown that the sequence of functions $\{\omega_m\}_{m\geq m_1}$ is monotonic and converges pointwise to the continuous function $\bar{w}$ on the compact interval $[0, T]$. Thus, by Dini's theorem [127, page 122], it follows that the convergence is uniform.

3.6 SIMULATION EXAMPLE

In this section a class I normalized Bouc–Wen model is considered, given by the following parameters:

$$\kappa_x = 2, \;\; \kappa_w = 2, \;\; \rho = 3, \;\; n = 1.5, \;\; \sigma = 1$$

A T-periodic triangular signal is chosen whose maximal amplitude is $X_{\max} = -X_{\min} = 0.2$. $T = 1$ and $T^{+} = T/2$ are taken. This signal is given in Figure 3.4 (upper left) on a one period of time.

Figure 3.4 (lower) gives two functions:

1. The function $\bar{\phi}_{\mathrm{BW}}(t)$ (solid) defined in Section 3.5.2 has been obtained using the analytical expressions (3.53) and (3.54) of the limit cycle.
2. The function $\Phi_{\mathrm{BW}}(x)(t)$ (dashed) has been obtained by solving the differential equations (3.9) and (3.10) of the normalized Bouc–Wen model with an initial condition $w(0) = 0$.

It can be seen that the output $\Phi_{\mathrm{BW}}(x)(t)$ approaches asymptotically the T-periodic function $\bar{\phi}_{\mathrm{BW}}(t)$ as predicted by Theorem 3.

Figure 3.4 (upper right) gives two functions:

1. The set of points $(x(t), \Phi_{BW}(x)(t))$ is obtained by solving the differential equations (3.9) and (3.10) of the normalized Bouc–Wen model (dashed).
2. The limit cycle is predicted by Equations (3.53) and (3.54) (solid).

It can be seen that the graph $(x(t), \Phi_{BW}(x)(t))$ approaches the limit cycle asymptotically as predicted by Theorem 3.

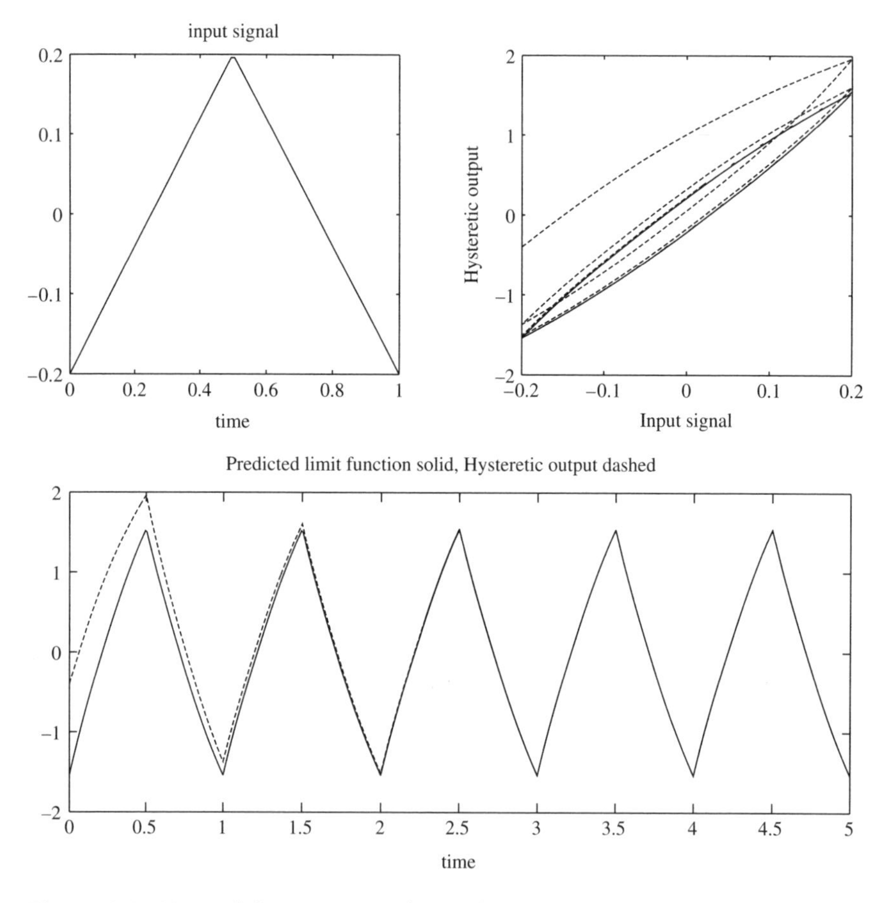

Figure 3.4 Upper left: input signal $x(\tau)$ for $0 \leq \tau \leq T$. Upper right: dashed, the graph $(x(t), \Phi_{BW}(t))$ for $t \in [0, 5T]$; solid, the graph of the limit cycle $(x(\tau), \bar{\Phi}_{BW}(\tau))$ for $0 \leq \tau \leq T$. Lower: dashed, the Bouc–Wen model output $\Phi_{BW}(x)(t)$; solid, the limit function $\bar{\phi}_{BW}(t)$ both for $t \in [0, 5T]$.

3.7 CONCLUSION

This chapter has dealt with the problem of characterizing analytically the periodic response of the Bouc–Wen model. Instead of using the standard form of this model, an equivalent form has been introduced: the normalized one. This normalized form of the Bouc–Wen model has been shown to be meaningful in the sense that the limit cycle depends directly on the parameters that appear in the normalized form, and thus depends only indirectly on the parameters of the standard form. The obtained expression of the limit cycle is explicit and exact and paves the way for a rigorous mathematical study of the relationship between the parameters of the model and the shape of the hysteretic cycle. This is the subject of the next chapter.

4

Variation of the Hysteresis Loop with the Bouc–Wen Model Parameters

4.1 INTRODUCTION

In this chapter, the hysteresis loop obtained for Bouc–Wen hysteresis is studied analytically. The normalized version of the model along with the analytical characterization of the limit cycle (Chapter 3) are the main tools for this study.

The relationship between the Bouc–Wen standard parameters A, β, γ, n, D, k, α and the shape of the hysteresis loop has been first studied in Reference [128]. In this reference, the value of $n = 1$ is considered, and the sign of the quantities $\beta + \gamma$ and $\gamma - \beta$ has been recognized to influence the general shape of the hysteresis loop. A more detailed study has been done in Reference [117] where five kinds of hysteresis loops have been related to the different combinations of the signs of $\beta + \gamma$ and $\gamma - \beta$. The parameter A has been shown to control the slope of the hysteresis at $z = 0$ and the parameter n has been recognized to control the transition from the linear to the nonlinear range. References [128] and [117] have also identified the quantity

$$z_0 = \sqrt[n]{\frac{A}{\beta + \gamma}}$$

Systems with Hysteresis: Analysis, Identification and Control using the Bouc–Wen Model
F. Ikhouane and J. Rodellar

as the largest value of the variable z. Other references include [129] and [130].

It is to be noted that until recently an analytical expression of the hysteresis loop (that is the limit cycle) was lacking, which has impeded an analytical study of the relationship between the the Bouc–Wen model parameters and the shape of the limit cycle. Most references dealing with this aspect have used numerical simulations to get more information (see [117] for example). However, numerical simulations can only give partial information as they use particular values of the model parameters. The objective of this chapter is to carry out an analytical study of the way the Bouc–Wen model parameters influence the shape of the hysteresis loop. This completes and improves the body of information available in the literature via numerical simulations.

The present chapter is an application of Theorem 3. An interesting property arising from this theorem is that the hysteresis loop depends *in a direct way* on a new set of parameters (the normalized ones). The objective of this chapter is to analyse how these parameters influence the limit cycle.

4.2 BACKGROUND RESULTS AND METHODOLOGY OF THE ANALYSIS

4.2.1 Background Results

In this chapter, it is considered that the input signal $x(t)$ is T-wave periodic (see Section 3.2.1). An example of such a signal is given in Figure 4.1. Without loss of generality, $X_{\max} = -X_{\min} > 0$ is taken so that it is possible to define a normalized signal

$$\bar{x} = \frac{x}{X_{\max}} \tag{4.1}$$

Note that in this case $-1 \leq \bar{x} \leq 1$.

For this class of inputs, Theorem 3 has been demonstrated in Chapter 3. This theorem is rewritten below to be used throughout this chapter.

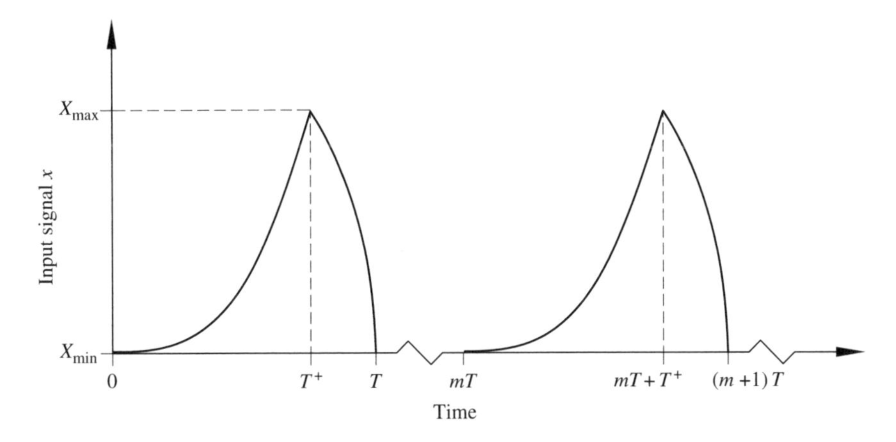

Figure 4.1 Example of a wave T-periodic signal.

Theorem 4. *For any nonnegative integer m, define the functions ω_m and ϕ_m as follows:*

$$\omega_m(\tau) = w(mT + \tau) \qquad \text{for } \tau \in [0, T] \tag{4.2}$$

$$\phi_m(\tau) = \kappa_x x(\tau) + \kappa_w \omega_m(\tau) \qquad \text{for } \tau \in [0, T] \tag{4.3}$$

Then the following can be found:

(*a*) *The sequences of functions $\{\phi_m\}_{m\geq 0}$ and $\{\omega_m\}_{m\geq 0}$ converge uniformly to the continuous functions $\bar{\Phi}_{\mathrm{BW}}$ and $\bar{w}$ defined in the form*

$$\bar{\Phi}_{\mathrm{BW}}(\tau) = \kappa_x X_{\max}\bar{x}(\tau) + \kappa_w \bar{w}(\tau) \qquad \text{for } \tau \in [0, T] \tag{4.4}$$

$$\bar{w}(\tau) = \psi^{+}_{\sigma,n}\left[\varphi^{+}_{\sigma,n}\left(-\psi_{\sigma,n}(\delta)\right) + \frac{\delta}{2}\left(\bar{x}(\tau) + 1\right)\right] \quad \text{for } \tau \in [0, T^+] \tag{4.5}$$

$$\bar{w}(\tau) = -\psi^{+}_{\sigma,n}\left[\varphi^{+}_{\sigma,n}\left(-\psi_{\sigma,n}(\delta)\right) - \frac{\delta}{2}\left(\bar{x}(\tau) - 1\right)\right] \quad \text{for } \tau \in [T^+, T] \tag{4.6}$$

where

$$\delta = 2\rho X_{\max} \tag{4.7}$$

(*b*) *For all* $\tau \in [0, T]$,

$$-1 < -\psi_{\sigma,n}(\delta) \leq \bar{w}(\tau) \leq \psi_{\sigma,n}(\delta) < 1 \tag{4.8}$$

the lower and upper bounds being attained at $\tau = 0$ *and* $\tau = T^+$ *respectively.*

Loosely speaking, Theorem 4 means that the hysteretic outputs $w(t)$ and $\Phi_{BW}(t)$ converge to T-periodic functions asymptotically. Since these limit functions are T-periodic, they need to be defined only on one period of time $[0, T]$. Theorem 4 gives the expression of the limit cycle that is obtained for the wave periodic input signal as an explicit function of the normalized Bouc–Wen model parameters.

The expression of the loading part of the limit cycle is given by the relations

$$\bar{\Phi}^{l}_{BW}(\bar{x}) = \kappa_x X_{max} \bar{x} + \kappa_w \bar{w}^{l}(\bar{x}) \tag{4.9}$$

$$\bar{w}^{l}(\bar{x}) = \psi^{+}_{\sigma,n}\left[\varphi^{+}_{\sigma,n}(-\psi_{\sigma,n}(\delta)) + \frac{\delta}{2}(\bar{x}+1)\right] \tag{4.10}$$

Similarly, the expression of the unloading part of the limit cycle is defined by the relations

$$\bar{\Phi}^{u}_{BW}(\bar{x}) = \kappa_x X_{max} \bar{x} + \kappa_w \bar{w}^{l}(\bar{x}) \tag{4.11}$$

$$\bar{w}^{u}(\bar{x}) = -\psi^{+}_{\sigma,n}\left[\varphi^{+}_{\sigma,n}(-\psi_{\sigma,n}(\delta)) - \frac{\delta}{2}(\bar{x}-1)\right] \tag{4.12}$$

where $-1 \leq \bar{x} \leq 1$.

From the expression of the limit cycle, it is clear that it depends in a direct way on the parameters σ, δ and n. This means that these parameters are those that shape directly the hysteresis loop and not the standard parameters A, β, γ, n, D whose influence on the loop is only indirect (via the set of parameters δ, σ, n).

Consider now two input values $\bar{x}_1$ and $\bar{x}_2 = -\bar{x}_1$. Then, from Equations (4.5) and (4.12),

$$\begin{aligned}\bar{w}^{l}(\bar{x}_1) &= \psi^{+}_{\sigma,n}\left[\varphi^{+}_{\sigma,n}(-\psi_{\sigma,n}(\delta)) + \frac{\delta}{2}(\bar{x}_1+1)\right] \\ &= \psi^{+}_{\sigma,n}\left[\varphi^{+}_{\sigma,n}(-\psi_{\sigma,n}(\delta)) - \frac{\delta}{2}(\bar{x}_2-1)\right] = -\bar{w}^{u}(\bar{x}_2)\end{aligned} \tag{4.13}$$

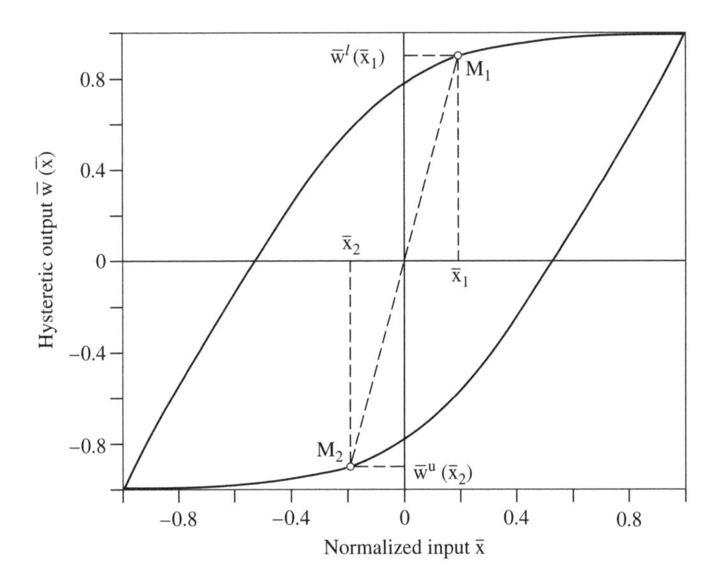

Figure 4.2 Symmetry property of the hysteresis loop of the Bouc–Wen model.

that is $\bar{w}^{\mathrm{l}}(\bar{x}_1) = -\bar{w}^{\mathrm{u}}(\bar{x}_2)$. This relation, along with Equations (4.9) and (4.11), leads to $\bar{\Phi}^{\mathrm{l}}_{\mathrm{BW}}(\bar{x}_1) = -\bar{\Phi}^{\mathrm{u}}_{\mathrm{BW}}(\bar{x}_2)$. This means that points M_1 and M_2 are symmetric with respect to the origin (see Figure 4.2). This symmetry property of the graph of the limit cycle allows only its loading or its unloading part to be used for analysing the effect of the parameters on the shape of the limit cycle. In the rest of the chapter, only the loading part of the limit cycle will be considered (unless otherwise specified). Equation (4.9) is linear in $\bar{x}$ and $\bar{w}$ so attention will be concentrated exclusively on Equation (4.5). The superscript 'l' for loading will be dropped for ease of notation. In the rest of the chapter it is considered that $n > 1$ and $\sigma > 1/2$ instead of $n \geq 1$ and $\sigma \geq 1/2$, so that the intervals for the analysis are open sets.

4.2.2 Methodology of the Analysis

To analyse the influence of the normalized set of parameters $\{\sigma, \delta, n\}$ on the shape of the limit cycle defined as the graph $(\bar{x}, \bar{w}(\bar{x}))$, three optics are considered:

1. The variations of the quantity $\bar{w}(\bar{x})$, seen as a function of each parameter σ, δ and n, are analysed separately for a fixed value of

the normalized input $\bar{x}$. This corresponds in Figure 4.3 to studying the evolution of the point Q along the $\bar{w}$ axis as the parameters σ, δ and n vary. This is done in Section 4.5.

The particular case $\bar{x} = 1$ corresponds to the maximal value of the hysteretic output. Due to the importance of this term, it will be studied separately in Section 4.3.

2. The second optics consists in an analysis along the $\bar{x}$ axis. The only point of interest in this axis is point S in Figure 4.3, which corresponds to $\bar{w}(\bar{x}) = 0$. This point gives the width of the hysteresis loop along the $\bar{x}$ axis. The analysis consists of studying the evolution of point S along the axis of abscissas when the parameters σ, δ and n vary. This is done in Section 4.4.
3. The third way in the analysis consists of defining four regions of the graph $(\bar{x}, \bar{w}(\bar{x}))$ as seen in Figure 4.3:

   (a) $R_l = [P_{sl}, \ P_{lt}]$, which corresponds to the linear behaviour.
   (b) $R_p = [P_{tp}, \ P_p]$, which corresponds to the plastic behaviour.
   (c) Two regions of transition $R_s = [P_s, \ P_{sl}]$ and $R_t = [P_{lt}, \ P_{tp}]$.

   The analysis consists of studying the evolution of the points that define each region. This is done in Section 4.6.

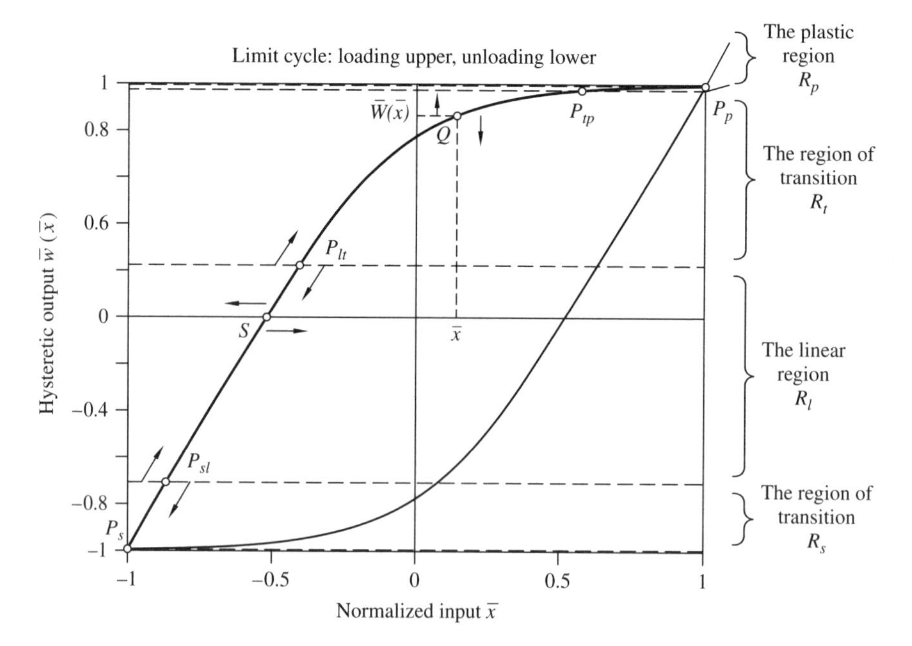

Figure 4.3 Methodologies of the analysis of the variation of $\bar{w}(\bar{x})$.

Note that Figure 4.3 is plotted in the following way:

1. On the axis of abscissas, the normalized input signal $\bar{x}$ is plotted, which varies between -1 and 1, both corresponding to the maximal amplitude (in absolute value) of the input signal $x(t)$.
2. On the axis of ordinates, there are in fact two plots: the upper one is that of the loading $\bar{w}^{l}(\bar{x})$ that corresponds to Equation (4.5) and the lower one corresponds to the unloading $\bar{w}^{u}(\bar{x})$ defined by Equation (4.12).

Since $\bar{w}(\cdot) \in (-1, 1)$ (see Theorem 4), this means that the plot is normalized both in the axis of abscissas and the axis of ordinates. The advantage of this normalization is that the shape of the hysteresis loop can be quantified in absolute terms while the use of the unnormalized variable z leads to hysteresis loops whose shape should always be reported to the maximal amplitude z_0 of the variable z, and to the maximal amplitude $X_{\max}$ of the input signal x.

4.3 MAXIMAL VALUE OF THE HYSTERETIC OUTPUT

By Theorem 4, the maximal value of the hysteretic output $\bar{w}(\bar{x})$ is attained for $\tau = T^{+}$ in Equation (4.6). Since $x(T^{+}) = X_{\max}$ (or equivalently $\bar{x}(T^{+}) = 1$), from Equation (4.6) it is found that $\bar{w}(T^{+}) = \psi_{\sigma,n}(\delta)$ as $\psi^{+}_{\sigma,n}$ is the inverse function of $\varphi^{+}_{\sigma,n}$. Using the notation of Equation (4.5), this means that the maximum value of the hysteretic output $\bar{w}(\bar{x})$ is $\psi_{\sigma,n}(\delta)$ and is obtained for the normalized input value $\bar{x} = 1$, that is $\bar{w}(1) = \psi_{\sigma,n}(\delta)$.

In this section the variation of the term $\psi_{\sigma,n}(\delta)$ is analysed with respect to each of the three parameters n, σ and δ.

4.3.1 Variation with Respect to δ

Since $\psi_{\sigma,n}$ is the inverse function of $\varphi_{\sigma,n}$,

$$\left[\frac{\partial \psi_{\sigma,n}(\mu)}{\partial \mu}\right]_{\mu=\delta} = \frac{1}{\left[\partial \varphi_{\sigma,n}(w)/\partial w\right]_{w=\psi_{\sigma,n}(\delta)}} \tag{4.14}$$

Note that, since $\delta > 0$, $w = \psi_{\sigma,n}(\delta) > 0$. Now it follows from Equations (3.18), (3.19) and (3.20) that

$$\frac{\partial \varphi_{\sigma,n}(w)}{\partial w} = \frac{1}{1+(2\sigma-1)w^n} + \frac{1}{1-w^n} \tag{4.15}$$

Combining Equations (4.14) and (4.15) gives

$$\left[\frac{\partial \psi_{\sigma,n}(\mu)}{\partial \mu}\right]_{\mu=\delta} = \frac{1}{[1/1+(2\sigma-1)\psi_{\sigma,n}(\delta)^n]+[1/1-\psi_{\sigma,n}(\delta)^n]} \tag{4.16}$$

By definition of the function $\psi_{\sigma,n}$, it is known that $\psi_{\sigma,n}(\delta) < 1$ and $\psi_{\sigma,n}(\delta) > 0$ as $\delta > 0$. Using these two inequalities in Equation (4.16), together with the fact that $\sigma > 1/2$, shows that

$$\left[\frac{\partial \psi_{\sigma,n}(\mu)}{\partial \mu}\right]_{\mu=\delta} > 0$$

This means that the function $\psi_{\sigma,n}(\delta)$ is increasing with δ.

On the other hand, from Equations (3.16) to (3.18), it follows that

$$\varphi_{\sigma,n}(0) = \varphi^+_{\sigma,n}(0) + \varphi^-_{\sigma,n}(0) = 0 \tag{4.17}$$

so that $\psi_{\sigma,n}(0) = 0$. Also, it has been shown in Section 3.4 that

$$\lim_{w\to 1} \varphi^+_{\sigma,n}(w) = +\infty$$

This implies by Equation (3.18) that

$$\lim_{w\to 1} \varphi_{\sigma,n}(w) = +\infty$$

so that

$$\lim_{\delta\to\infty} \psi_{\sigma,n}(\delta) = 1 \tag{4.18}$$

Figure 4.4 gives a plot of a function $\psi_{\sigma,n}$.

The property (4.18) means that, by increasing the parameter δ, the maximal value of the hysteretic output reaches unity asymptotically. Let us now interpret this result in terms of the unnormalized model (2.4)–(2.5). As seen in Theorem 4, the hysteretic variable z

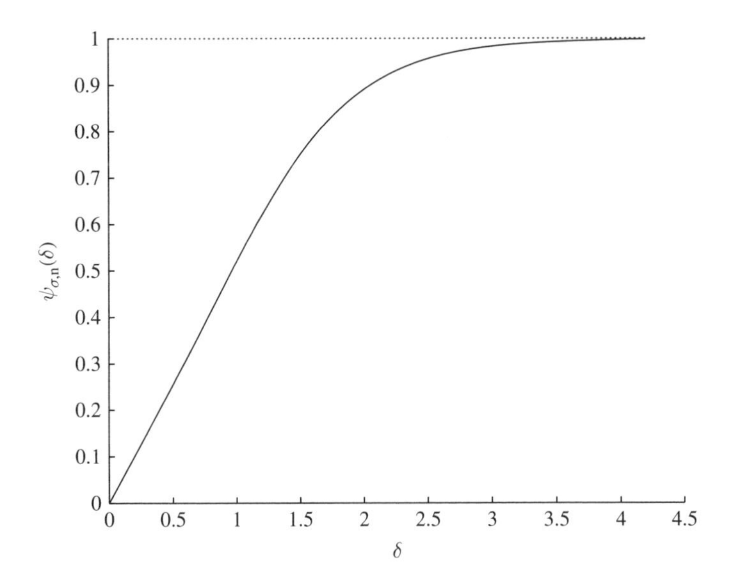

Figure 4.4 Variation of the maximal hysteretic output $\psi_{\sigma,n}(\delta)$ with the parameter δ, for the values of $\sigma = 2$ and $n = 2$.

reaches a stationary state when the input signal x is wave periodic. The obtained limit cycle (x, z) then has a maximal value for

$$z = z_{\max} = \psi_{\sigma,n}(\delta)\, z_0 \qquad \text{and} \qquad x = X_{\max} \qquad \text{where} \quad z_0 = \sqrt[n]{\frac{A}{\beta+\gamma}}$$

This maximal amplitude $z_{\max}$ of the hysteretic term z may be substantially different from z_0 as the term $\psi_{\sigma,n}(\delta)$ can take any value between 0 and 1 for different values of the Bouc–Wen model parameters. However, the quantity $z_{\max}$ goes asymptotically to z_0 as the parameter δ goes to infinity.

4.3.2 Variation with Respect to σ

The function $\psi_{\sigma,n}(\mu)$ may be seen as a function ψ of the three variables σ, n and μ. Similarly, the function $\varphi_{\sigma,n}(w)$ may be considered as a function φ of σ, n and w. More precisely these functions have the following structure:

$$\psi: \mathbb{R}^3 \rightarrow \mathbb{R} \qquad\qquad \varphi: \mathbb{R}^3 \rightarrow \mathbb{R}$$

$$\begin{pmatrix} \sigma \\ n \\ \mu \end{pmatrix} \rightarrow \psi_{\sigma,n}(\mu) \quad \text{and} \quad \begin{pmatrix} \sigma \\ n \\ w \end{pmatrix} \rightarrow \varphi_{\sigma,n}(w)$$

with the property that

$$\varphi_{\sigma,n}\left(\psi_{\sigma,n}(\mu)\right) = \mu \qquad \text{for all } \sigma > 1/2,\ n > 1,\ \mu \in \mathbb{R} \tag{4.19}$$

Define the function h as follows:

$$h: \mathbb{R}^3 \rightarrow \mathbb{R}^3$$
$$\begin{pmatrix} \sigma \\ n \\ \mu \end{pmatrix} \rightarrow \begin{pmatrix} \sigma \\ n \\ \psi_{\sigma,n}(\mu) \end{pmatrix}$$

Then, Equation (4.19) can be written as

$$\varphi\left[h(\sigma, n, \mu)^{\mathrm{T}}\right] = \mu \tag{4.20}$$

It follows from Equation (4.20) that

$$J_\varphi J_h = (0, 0, 1) \tag{4.21}$$

where J_φ is the Jacobian matrix of the function φ evaluated at the point $h(\sigma, n, \mu)^{\mathrm{T}}$ and J_h is the Jacobian matrix of the function h evaluated at the point $(\sigma, n, \mu)^{\mathrm{T}}$. Equation (4.21) can be written more explicitly as

$$\left(\frac{\partial \varphi_{\sigma,n}(w)}{\partial \sigma}, \frac{\partial \varphi_{\sigma,n}(w)}{\partial n}, \frac{\partial \varphi_{\sigma,n}(w)}{\partial w} \right)_{w=\psi_{\sigma,n}(\mu)}$$

$$\times \begin{pmatrix} 1 & 0 & \dfrac{\partial \psi_{\sigma,n}(\mu)}{\partial \sigma} \\ 0 & 1 & \dfrac{\partial \psi_{\sigma,n}(\mu)}{\partial n} \\ \dfrac{\partial \psi_{\sigma,n}(\mu)}{\partial \sigma} & \dfrac{\partial \psi_{\sigma,n}(\mu)}{\partial n} & \dfrac{\partial \psi_{\sigma,n}(\mu)}{\partial \mu} \end{pmatrix} = (0, 0, 1) \tag{4.22}$$

Equation (4.22) leads to

$$\left[\frac{\partial\psi_{\sigma,n}(\mu)}{\partial\sigma}\right]_{\mu=\delta} = -\frac{\left[\dfrac{\partial\varphi_{\sigma,n}(w)}{\partial\sigma}\right]_{w=\psi_{\sigma,n}(\delta)}}{\left[\dfrac{\partial\varphi_{\sigma,n}(w)}{\partial w}\right]_{w=\psi_{\sigma,n}(\delta)}} \tag{4.23}$$

From Equation (3.21),

$$\begin{aligned}\left[\frac{\partial\varphi_{\sigma,n}(w)}{\partial\sigma}\right]_{w=\psi_{\sigma,n}(\delta)} &= \left[\int_0^w \frac{\partial}{\partial\sigma}\left[\frac{1}{1+(2\sigma-1)u^n}\right]\mathrm{d}u\right]_{w=\psi_{\sigma,n}(\delta)} \\ &= -\int_0^{\psi_{\sigma,n}(\delta)} \frac{2u^n}{[1+(2\sigma-1)u^n]^2}\mathrm{d}u \end{aligned} \tag{4.24}$$

and

$$\left[\frac{\partial\varphi_{\sigma,n}(w)}{\partial w}\right]_{w=\psi_{\sigma,n}(\delta)} = \frac{1}{1-\psi_{\sigma,n}(\delta)^n} + \frac{1}{1+(2\sigma-1)\psi_{\sigma,n}(\delta)^n} \tag{4.25}$$

Note that, in Equations (4.24) and (4.25), $0 < \psi_{\sigma,n}(\delta) < 1$ and $\sigma > 1/2$, so that

$$\left[\frac{\partial\varphi_{\sigma,n}(w)}{\partial\sigma}\right]_{w=\psi_{\sigma,n}(\delta)} < 0 \qquad \text{and} \qquad \left[\frac{\partial\varphi_{\sigma,n}(w)}{\partial w}\right]_{w=\psi_{\sigma,n}(\delta)} > 0$$

This means by Equation (4.23) that $\partial\psi_{\sigma,n}(\delta)/\partial\sigma > 0$ so the function $\psi_{\sigma,n}(\delta)$ is increasing with the argument σ. The value of the function $\psi_{\sigma,n}(\delta)$ for $\sigma = 1/2$ is $\psi_{1/2,n}(\delta)$. To determine the asymptotic behaviour of the function $\psi_{\sigma,n}(\delta)$ the behaviour of the function $\varphi_{\sigma,n}(w)$ has to be analysed as $\sigma \to \infty$. Applying Lebesgue's monotone convergence theorem [131, page 21] it follows from Equation (3.19) that

$$\lim_{\sigma\to\infty} \varphi^-_{\sigma,n}(w) = 0 \qquad \text{for all } 0 \le w \le 1 \text{ and } n > 1$$

On the other hand, using the notation of Equation (3.35) it is found that, for all $0 \le w < 1$, $\varphi_{\sigma,n}(w) = \varphi^-_{\sigma,n}(w) + \varphi^+_n(w)$ so that

$$\lim_{\sigma\to\infty} \varphi_{\sigma,n}(w) = \lim_{\sigma\to\infty} \left[\varphi^-_{\sigma,n}(w) + \varphi^+_n(w)\right] = \varphi^+_n(w)$$

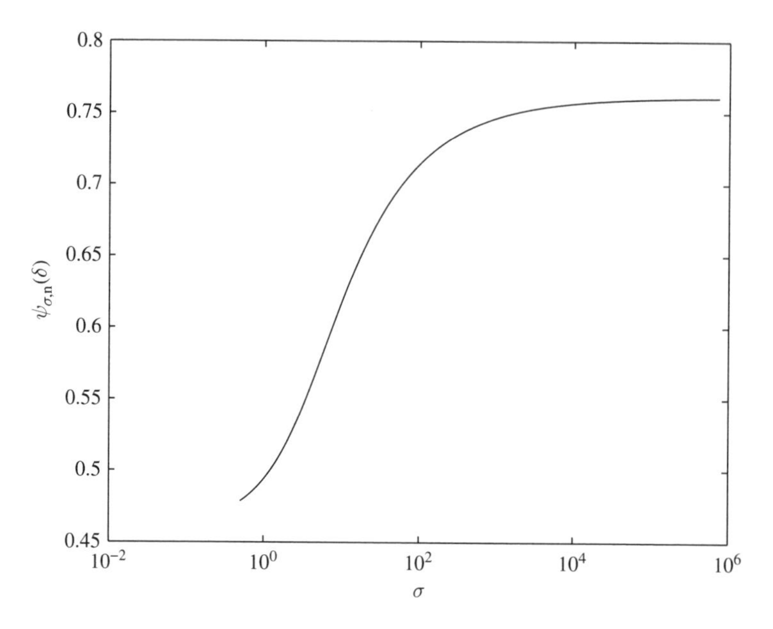

Figure 4.5 Variation of the maximal hysteretic output $\psi_{\sigma,n}(\delta)$ with the parameter σ, for the values of $\delta = 1$ and $n = 2$ (semi-logarithmic scale). Observe that for $\sigma = 0.5$ the corresponding value is $\psi_{0.5,2}(1) = 0.4786$ and $\lim_{\sigma\to\infty} \psi_{\sigma,2}(1) = 0.7610 = \psi_2^+(1)$.

This means that

$$\lim_{\sigma\to\infty} \psi_{\sigma,n}(\delta) = \psi_n^+(\delta)$$

Figure 4.5 gives the evolution of the function $\psi_{\sigma,n}(\delta)$ with the parameter σ.

4.3.3 Variation with Respect to *n*

From Equation (4.22),

$$\left[\frac{\partial \psi_{\sigma,n}(\mu)}{\partial n}\right]_{\mu=\delta} = -\frac{\left[\dfrac{\partial \varphi_{\sigma,n}(w)}{\partial n}\right]_{w=\psi_{\sigma,n}(\delta)}}{\left[\dfrac{\partial \varphi_{\sigma,n}(w)}{\partial w}\right]_{w=\psi_{\sigma,n}(\delta)}} \tag{4.26}$$

Equation (3.21) gives

$$\left[\frac{\partial \varphi_{\sigma,n}(w)}{\partial n}\right]_{w=\psi_{\sigma,n}(\delta)} = \left[\int_0^w \frac{\partial}{\partial n}\left[\frac{1}{1+(2\sigma-1)u^n}\right] du\right]_{w=\psi_{\sigma,n}(\delta)}$$
$$+\left[\int_0^w \frac{\partial}{\partial n}\left(\frac{1}{1-u^n}\right) du\right]_{w=\psi_{\sigma,n}(\delta)}$$
$$= \int_0^{\psi_{\sigma,n}(\delta)} \left[\frac{1}{(1-u^n)^2} - \frac{2\sigma-1}{(1+(2\sigma-1)u^n)^2}\right] \log(u)u^n du \tag{4.27}$$

In order to analyse the sign of the right-hand term of Equation (4.27), two cases are to be discussed: $\sigma \leq 1$ and $\sigma > 1$.

The Case $\sigma \leq 1$

It can be seen that the right-hand term of Equation (4.27) is negative as $0 < \psi_{\sigma,n}(\delta) < 1$. Since, by Equation (4.25),

$$\left[\frac{\partial \varphi_{\sigma,n}(w)}{\partial w}\right]_{w=\psi_{\sigma,n}(\delta)} > 0$$

it follows that

$$\left[\frac{\partial \psi_{\sigma,n}(\mu)}{\partial n}\right]_{\mu=\delta} > 0$$

by Equation (4.26). This means that the function $\psi_{\sigma,n}(\delta)$ is increasing with the variable n. The value of this function for $n = 1$ is $\psi_{\sigma,1}(\delta)$. To analyse the asymptotic behaviour of $\psi_{\sigma,n}(\delta)$ the following lemma is needed.

Lemma 8. *For all* $\sigma > 1/2$,

$$\lim_{n\to\infty} \psi^+_{\sigma,n}(\mu) = \mu \qquad \textit{for } \mu \in [-1, 1) \tag{4.28}$$

$$\lim_{n\to\infty} \psi^+_{\sigma,n}(\mu) = 1 \qquad \textit{for } \mu \geq 1 \tag{4.29}$$

$$\lim_{n\to\infty} \psi^-_{\sigma,n}(\mu) = \mu \qquad \textit{for } \mu \in (-1, 1] \tag{4.30}$$

$$\lim_{n\to\infty} \psi_{\sigma,n}(\mu) = \frac{\mu}{2} \qquad \textit{for } \mu \in (-2, 2] \tag{4.31}$$

$$\lim_{n\to\infty} \psi_{\sigma,n}(\mu) = 1 \qquad \textit{for } \mu > 2 \tag{4.32}$$

Proof. Using the Lebesgue monotone convergence theorem on any interval $[-k, k]$ with $0 < k < 1$, it follows from Equations (3.16) to (3.18) that

$$\lim_{n\to\infty} \varphi^{+}_{\sigma,n}(w) = w \qquad \text{for } w \in [-1, 1) \tag{4.33}$$

$$\lim_{n\to\infty} \varphi^{-}_{\sigma,n}(w) = w \qquad \text{for } w \in (-1, 1] \tag{4.34}$$

$$\lim_{n\to\infty} \varphi_{\sigma,n}(w) = 2w \qquad \text{for } w \in (-1, 1) \tag{4.35}$$

Equations (4.28), (4.30) and (4.31) follow from Equations (4.33), (4.34) and (4.35) respectively. To prove Equation (4.29), take $\mu \geq 1$ and denote $w_k = \psi^{+}_{\sigma,k}(\mu)$ for any positive integer k. It is required that the sequence $\{w_k\}$ is shown to be increasing. To this end note that, similar to Equation (4.26),

$$\frac{\partial \psi^{+}_{\sigma,n}(\mu)}{\partial n} = -\frac{\left[\dfrac{\partial \varphi^{+}_{\sigma,n}(w)}{\partial n}\right]_{w=\psi^{+}_{\sigma,n}(\mu)}}{\left[\dfrac{\partial \varphi^{+}_{\sigma,n}(w)}{\partial w}\right]_{w=\psi^{+}_{\sigma,n}(\mu)}} \tag{4.36}$$

From Equation (3.17) it follows that

$$\left[\frac{\partial \varphi^{+}_{\sigma,n}(w)}{\partial n}\right]_{w=\psi^{+}_{\sigma,n}(\mu)} = \int_{0}^{\psi_{\sigma,n}(\delta)} \frac{\log(u)u^{n}}{(1-u^{n})^{2}} du < 0 \tag{4.37}$$

$$\left[\frac{\partial \varphi^{+}_{\sigma,n}(w)}{\partial w}\right]_{w=\psi^{+}_{\sigma,n}(\mu)} = \frac{1}{1-\psi^{+}_{\sigma,n}(\mu)^{n}} > 0 \tag{4.38}$$

The fact that the function $\psi^{+}_{\sigma,n}(\mu)$ is increasing with n is a direct consequence of Equations (4.36) to (4.38).

Now, it has been proved that the sequence $\{w_k\}$ is increasing. Since $w_k < 1$ for all integers k, it follows that a real number $\ell \leq 1$ exists such that

$$\lim_{k\to\infty} w_k = \ell$$

Assume that $\ell < 1$ and, for any positive integers k, then $w_k \leq \ell$, or equivalently $\psi^+_{\sigma,k}(\mu) \leq \ell$. This implies that $\mu \leq \varphi^+_{\sigma,k}(\ell)$ for all positive integers k. Using Equation (4.33) (as $0 \leq \ell < 1$) it follows that $\mu \leq \ell$. This contradicts the fact that if $\mu \geq 1$ then necessarily $\ell = 1$. In other terms, it has been shown that

$$\lim_{k\to\infty} w_k = \lim_{k\to\infty} \psi^+_{\sigma,k}(\mu) = \ell = 1$$

which demonstrates Equation (4.29). Equation (4.32) follows using similar arguments.

Lemma 8 shows that

$$\lim_{n\to\infty} \psi_{\sigma,n}(\delta) = \frac{\delta}{2} \quad \text{for } \delta \in (0,2] \quad \text{and} \quad \lim_{n\to\infty} \psi_{\sigma,n}(\delta) = 1 \quad \text{for } \delta > 2$$

Figure 4.6 gives the variation of the function $\psi_{\sigma,n}(\delta)$ for $\sigma = 0.7 \leq 1$.

The Case $\sigma > 1$

When $\sigma > 1$, an analytical description of the variation of $\psi_{\sigma,n}(\delta)$ is difficult. Numerical simulations show that a value $\sigma^* > 1$ exists that depends on the parameter δ, such that

1. If $\sigma \leq \sigma^*$, then $\psi_{\sigma,n}(\delta)$ is increasing with n.
2. If $\sigma > \sigma^*$, then the function $\psi_{\sigma,n}(\delta)$ increases with n, attains a maximum and then decreases as $n \to \infty$ (see Figure 4.6).

In this case

$$\lim_{n\to\infty} \psi_{\sigma,n}(\delta) = \frac{\delta}{2} \quad \text{for } \delta \in (0,2] \quad \text{and} \quad \lim_{n\to\infty} \psi_{\sigma,n}(\delta) = 1 \quad \text{for } \delta > 2$$

Figure 4.6 gives the variation of the maximal value $\psi_{\sigma,n}(\delta)$ with the parameter n for three values of σ and with $\delta = 1.4$.

4.3.4 Summary of the Obtained Results

Table 4.1 summarizes the obtained results. For example, the second row of the table should be read as follows:

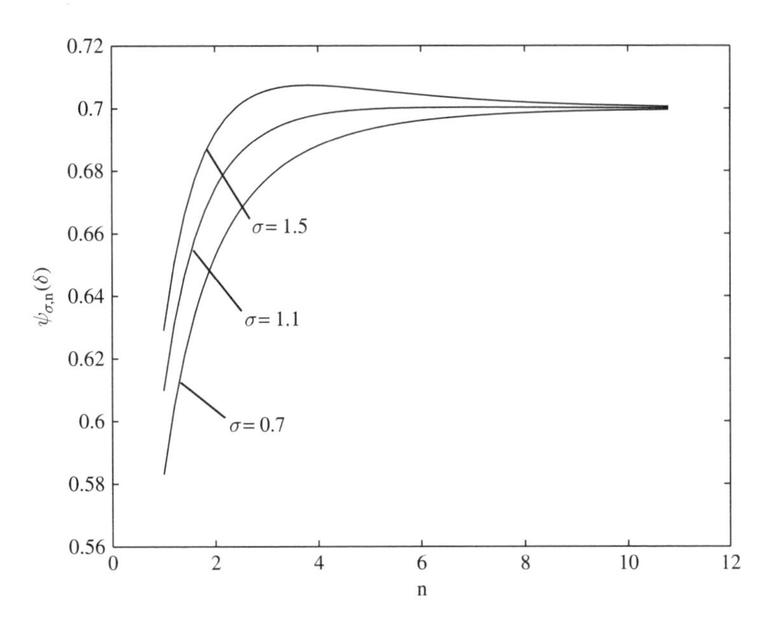

Figure 4.6 Variation of the maximal value $\psi_{\sigma,n}(\delta)$ with the parameter n for three values of σ and with $\delta = 1.4$. In this case, we have $\sigma^* \simeq 1.1$.

Table 4.1 Variation of the maximal hysteretic output $\psi_{\sigma,n}(\delta)$ with the Bouc–Wen model parameters δ, σ, n

$\psi_{\sigma,n}(\delta)$	0	↑	1
δ	0		$+\infty$
$\psi_{\sigma,n}(\delta)$	$\psi_{1/2,n}(\delta)$	↑	$\psi_n^+(\delta)$
σ	1/2		$+\infty$
$\psi_{\sigma,n}(\delta)$	$\psi_{\sigma,1}(\delta)$	↑	$\begin{cases} \delta/2 & \text{if } \delta \in (0,2] \\ 1 & \text{if } \delta > 2 \end{cases}$
n with $\sigma \leq 1$	1		$+\infty$
$\psi_{\sigma,n}(\delta)$	$\psi_{\sigma,1}(\delta)$		$\begin{cases} \delta/2 & \text{if } \delta \in (0,2] \\ 1 & \text{if } \delta > 2 \end{cases}$
n with $\sigma > 1$	1		$+\infty$

- The first column shows the variation of $\psi_{\sigma,n}(\delta)$ with respect to the parameter σ.
- The second column shows that the term $\psi_{\sigma,n}(\delta)$ is increasing (↑) from the value $\psi_{1/2,n}(\delta)$ that corresponds to $\sigma = 1/2$ to the value $\psi_n^+(\delta)$ that corresponds to $\sigma = +\infty$.

In Table 4.1 (and in all other similar tables), only the results that have been derived *analytically* are reported. For example, it has not been possible to derive analytically the growth of $\psi_{\sigma,n}(\delta)$ (that is whether this term is increasing ($\uparrow$) or decreasing ($\downarrow$)) with respect to n when $\sigma > 1$ (fourth line of Table 4.1) so this growth has not been reported in the table.

4.4 VARIATION OF THE ZERO OF THE HYSTERETIC OUTPUT

In this section the variation of the normalized input value $\bar{x}^\circ$ is analysed such that $\bar{w}(\bar{x}^\circ) = 0$. The value $\bar{x}^\circ$ corresponds to the abscissa of the point S in Figure 4.3. Using Equation (4.5) it follows that

$$\bar{x}^\circ = -\frac{2}{\delta}\varphi^{+}_{\sigma,n}\left[-\psi_{\sigma,n}(\delta)\right] - 1 = \frac{2}{\delta}\varphi^{-}_{\sigma,n}\left[\psi_{\sigma,n}(\delta)\right] - 1 \tag{4.39}$$

Since $\varphi^{-}_{\sigma,n}(w) < \varphi^{+}_{\sigma,n}(w)$ for $w \neq 0$, then $\varphi^{-}_{\sigma,n}(w) < \varphi_{\sigma,n}(w)/2$ for $w \neq 0$ so that, taking $w = \psi_{\sigma,n}(\delta)$, this becomes $\bar{x}^\circ < 0$. Note that the quantity $|2\bar{x}^\circ|$ corresponds to the width of the hysteresis loop along the $\bar{x}$ axis due to the symmetry property of the limit cycle.

4.4.1 Variation with Respect to δ

Equation (4.39) gives the following derivative:

$$\frac{\partial \bar{x}^\circ}{\partial \delta} = \frac{2}{\delta^2}\left\{\frac{\delta}{1 + \dfrac{1 + (2\sigma - 1)\psi_{\sigma,n}(\delta)^n}{1 - \psi_{\sigma,n}(\delta)^n}} + \varphi^{+}_{\sigma,n}\left[-\psi_{\sigma,n}(\delta)\right]\right\} \tag{4.40}$$

The objective of the subsequent analysis is to determine the sign of the right-hand term in Equation (4.39). Combining Equations (4.16) and (3.23) this becomes

$$\frac{\partial \varphi^{+}_{\sigma,n}\left[-\psi_{\sigma,n}(\delta)\right]}{\partial \delta} = -\frac{\partial \psi_{\sigma,n}(\delta)}{\partial \delta}\left[\frac{\partial \varphi^{+}_{\sigma,n}(w)}{\partial w}\right]_{w=-\psi_{\sigma,n}(\delta)}$$

$$= -\frac{1}{1+\dfrac{1+(2\sigma-1)\psi_{\sigma,n}(\delta)^n}{1-\psi_{\sigma,n}(\delta)^n}} \tag{4.41}$$

Define the function

$$f(\delta) = \delta + \varphi^+_{\sigma,n}\left[-\psi_{\sigma,n}(\delta)\right]\left[1+\frac{1+(2\sigma-1)\psi_{\sigma,n}(\delta)^n}{1-\psi_{\sigma,n}(\delta)^n}\right] \tag{4.42}$$

From Equations (4.41) and (4.16) it follows that

$$\frac{\mathrm{d}f(\delta)}{\mathrm{d}\delta} = \frac{2\sigma n\varphi^+_{\sigma,n}\left[-\psi_{\sigma,n}(\delta)\right]\psi_{\sigma,n}(\delta)^n}{\left[1-\psi_{\sigma,n}(\delta)^n\right]^2\left[\dfrac{1}{1+(2\sigma-1)\psi_{\sigma,n}(\delta)^n}+\dfrac{1}{1-\psi_{\sigma,n}(\delta)^n}\right]} \tag{4.43}$$

From Equation (4.43), it is clear that

$$\frac{\mathrm{d}f(\delta)}{\mathrm{d}\delta} < 0$$

Since $f(0) = 0$ by Equation (4.42), this means that $f(\delta) < 0$ for all $\delta > 0$. Combining Equations (4.40) and (4.42) gives

$$\frac{\partial \bar{x}^\circ}{\partial \delta} = \frac{2f(\delta)}{\delta^2\left[1+\dfrac{1+(2\sigma-1)\psi_{\sigma,n}(\delta)^n}{1-\psi_{\sigma,n}(\delta)^n}\right]} \tag{4.44}$$

Equation (4.44) shows that

$$\frac{\partial \bar{x}^\circ}{\partial \delta} < 0 \text{ for all } \delta > 0$$

so that the function $\bar{x}^\circ$ is decreasing with the parameter δ. The value of $\bar{x}^\circ$ at $\delta = 0$ is determined as

$$\bar{x}^\circ(\delta = 0) = \lim_{\delta\to 0}\frac{2}{\delta}\varphi^-_{\sigma,n}\left[\psi_{\sigma,n}(\delta)\right] - 1 = 2\left(\frac{\partial\varphi^-_{\sigma,n}\left[\psi_{\sigma,n}(\delta)\right]}{\partial\delta}\right)_{\delta=0} - 1 = 0 \tag{4.45}$$

Using the fact that

$$\lim_{\delta\to\infty}\psi_{\sigma,n}(\delta) = 1$$

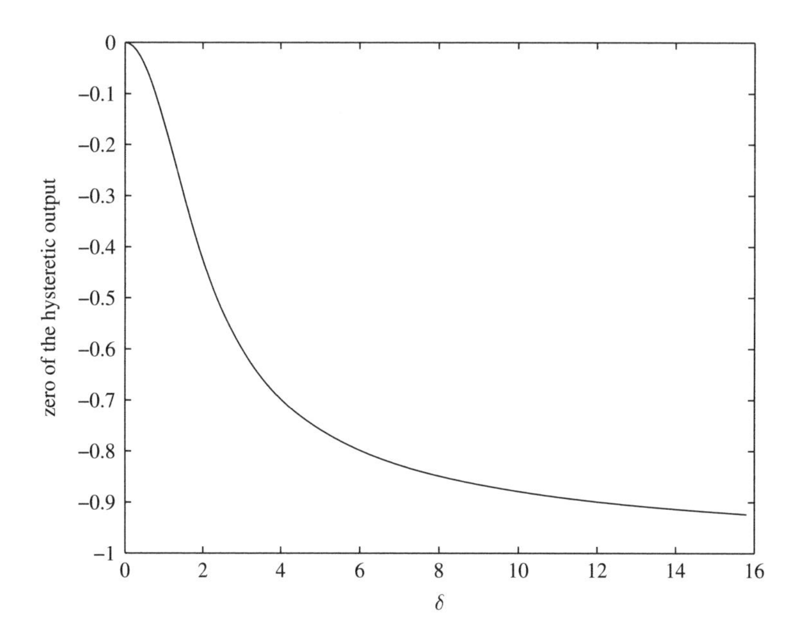

Figure 4.7 Variation of $\bar{x}^{\circ}$ with the parameter δ with the values $\sigma = 2$ and $n = 2$.

and that the function $\varphi^{-}_{\sigma,n}(\mu)$ is finite at $\mu = 1$, from equation (4.39) it follows that

$$\lim_{\delta\to\infty} \bar{x}^{\circ}(\delta) = -1$$

Figure 4.7 gives an example of the variation of $\bar{x}^{\circ}$ with the parameter δ for some given values of the parameters σ and n.

4.4.2 Variation with Respect to σ

To compute the derivative of the term $\bar{x}^{\circ}$ with respect to σ, a technique similar to that of Section 4.3.2 is used. Using the composition rule in Equation (4.39) gives

$$\frac{\partial \bar{x}^{\circ}}{\partial \sigma} = -\frac{2}{\delta}\left\{\left[\frac{\partial \varphi^{+}_{\sigma,n}(w)}{\partial \sigma}\right]_{w=-\psi_{\sigma,n}(\delta)} - \left[\frac{\partial \varphi^{+}_{\sigma,n}(w)}{\partial w}\right]_{w=-\psi_{\sigma,n}(\delta)} \left[\frac{\partial \psi_{\sigma,n}(\mu)}{\partial \sigma}\right]_{\mu=\delta}\right\} \tag{4.46}$$

Then the terms arising in Equation (4.46) are computed:

$$\left[\frac{\partial \varphi^{+}_{\sigma,n}(w)}{\partial \sigma}\right]_{w=-\psi_{\sigma,n}(\delta)} = \int_{0}^{\psi_{\sigma,n}(\delta)} \frac{2u^{n}}{[1+(2\sigma-1)u^{n}]^{2}}\mathrm{d}u \tag{4.47}$$

$$\left[\frac{\partial \varphi^{+}_{\sigma,n}(w)}{\partial w}\right]_{w=-\psi_{\sigma,n}(\delta)} = \frac{1}{1+(2\sigma-1)\psi_{\sigma,n}(\delta)^{n}} \tag{4.48}$$

Equations (4.46) to (4.48) along with Equations (4.23) to (4.25) give

$$\frac{\partial \bar{x}^{\circ}}{\partial \sigma} = -\frac{2}{\delta}\left[1 - \frac{1}{1+\dfrac{1+(2\sigma-1)\psi_{\sigma,n}(\delta)^{n}}{1-\psi_{\sigma,n}(\delta)^{n}}}\right] \times \int_{0}^{\psi_{\sigma,n}(\delta)} \frac{2u^{n}}{[1+(2\sigma-1)u^{n}]^{2}}\mathrm{d}u < 0 \tag{4.49}$$

Equation (4.49) implies that $\bar{x}^{\circ}$ is a decreasing function of the parameter σ. It can easily be checked that $\bar{x}^{\circ}$ decreases from the value

$$\frac{2}{\delta}\psi_{1/2,n}(\delta) - 1$$

at $\sigma = 1/2$ to -1 when $\sigma \to \infty$.

Figure 4.8 gives an example of the variation of $\bar{x}^{\circ}$ with the parameter σ for some given values of the parameters δ and n.

4.4.3 Variation with Respect to n

From Equation (4.39),

$$\frac{\partial \bar{x}^{\circ}}{\partial n} = -\frac{2}{\delta}\left\{\left[\frac{\partial \varphi^{+}_{\sigma,n}(w)}{\partial n}\right]_{w=-\psi_{\sigma,n}(\delta)} - \left(\frac{\partial \varphi^{+}_{\sigma,n}(w)}{\partial w}\right)_{w=-\psi_{\sigma,n}(\delta)}\left[\frac{\partial \psi_{\sigma,n}(\mu)}{\partial n}\right]_{\mu=\delta}\right\} \tag{4.50}$$

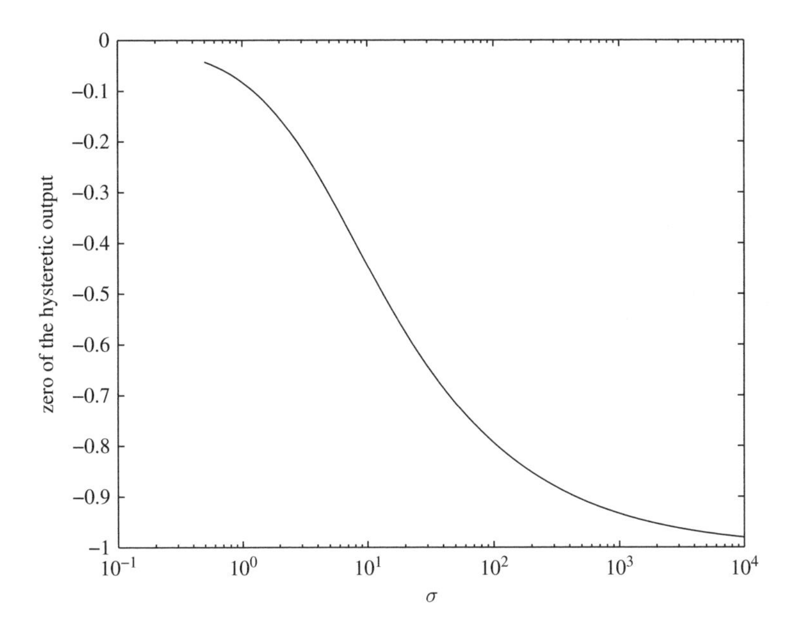

Figure 4.8 Variation of $\bar{x}^{\circ}$ with the parameter σ with the values $\delta = 1$ and $n = 2$.

On the other hand,

$$\left[\frac{\partial \varphi^{+}_{\sigma,n}(w)}{\partial n}\right]_{w=-\psi_{\sigma,n}(\delta)} = \int_0^{\psi_{\sigma,n}(\delta)} \frac{(2\sigma-1)\log(u)u^n}{[1+(2\sigma-1)u^n]^2}\mathrm{d}u \tag{4.51}$$

Combining Equations (4.50) and (4.51) along with (4.48) and (4.25) to (4.27) gives

$$\frac{\partial \bar{x}^{\circ}}{\partial n} = -\frac{2}{\delta}\left\{(1-\Gamma)\int_0^{\psi_{\sigma,n}(\delta)} \frac{(2\sigma-1)\log(u)u^n}{[1+(2\sigma-1)u^n]^2}\mathrm{d}u + \Gamma\int_0^{\psi_{\sigma,n}(\delta)} \frac{\log(u)u^n}{(1-u^n)^2}\mathrm{d}u\right\} > 0 \tag{4.52}$$

where

$$\Gamma = \frac{1}{1+\dfrac{1+(2\sigma-1)\psi_{\sigma,n}(\delta)^n}{1-\psi_{\sigma,n}(\delta)^n}}$$

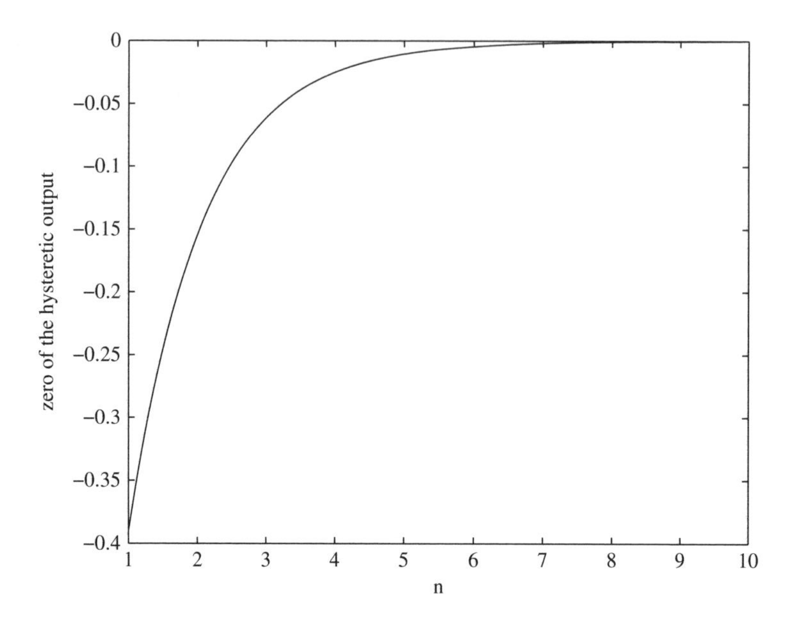

Figure 4.9 Variation of $\bar{x}^{\circ}$ with the parameter n with the values $\delta = 1$ and $\sigma = 2$.

Equation (4.52) shows that the term $\bar{x}^{\circ}$ is increasing with the parameter n. At $n = 1$, the value of $\bar{x}^{\circ}$ is given by taking $n = 1$ in Equation (4.39). Since $\bar{x}^{\circ} < 0$ for all $n \geq 1$ and since it is increasing with n by Equation (4.52) it goes to a finite limit. Lemma 8 shows that

$$\lim_{n\to\infty} \bar{x}^{\circ} = 0 \qquad \text{for } 0 < \delta \leq 2 \quad \text{and} \quad \lim_{n\to\infty} \bar{x}^{\circ} = \frac{2}{\delta} - 1 \qquad \text{for } \delta > 2$$

Figure 4.9 gives an example of the variation of $\bar{x}^{\circ}$ with the parameter n for some given values of the parameters δ and σ.

4.4.4 Summary of the Obtained Results

This section summarizes the obtained results in Table 4.2. For example, the third row gives the variation of $\bar{x}^{\circ}$ with the parameter n in the following form:

Table 4.2 Variation of the hysteretic zero $\bar{x}^\circ$ with the Bouc–Wen model parameters δ, σ, n

$\bar{x}^\circ$	0	$\downarrow$	-1
δ	0		$+\infty$
$\bar{x}^\circ$	$(2/\delta)\psi_{1/2,n}(\delta)-1$	$\downarrow$	-1
σ	$1/2$		$+\infty$
$\bar{x}^\circ$	$(2/\delta)\varphi^-_{\sigma,1}\left[\psi_{\sigma,1}(\delta)\right]-1$	$\uparrow$	$\begin{cases}0 & \text{if } \delta\in(0,2]\\ 2/\delta-1 & \text{if } \delta>2\end{cases}$
n	1		$+\infty$

- For $n=1$, the corresponding value of $\bar{x}^\circ$ is $(2/\delta)\varphi^-_{\sigma,1}\left[\psi_{\sigma,1}(\delta)\right]-1$.
- The symbol $\uparrow$ indicates that $\bar{x}^\circ$ is increasing with n. Also, this line indicates that

$$\lim_{n\to\infty}\bar{x}^\circ=0 \quad \text{if } \delta\in(0,2] \quad \text{and} \quad \lim_{n\to\infty}\bar{x}^\circ=\frac{2}{\delta}-1 \quad \text{if } \delta>2$$

4.5 VARIATION OF THE HYSTERETIC OUTPUT WITH THE BOUC–WEN MODEL PARAMETERS

In this section the quantity $\bar{w}(\bar{x})$ of Equation (4.5) is considered as a function of the parameters δ, σ and n with a fixed value of $\bar{x}$. This corresponds to studying the evolution of the point Q along the axis of ordinates in Figure 4.3. Seen as a function of $\bar{x}$ with fixed parameters δ, σ and n, the function $\bar{w}(\bar{x})$ is strictly increasing. This can easily be seen by computing its derivative from Equation (4.5):

$$\frac{d\bar{w}(\bar{x})}{d\bar{x}}=\frac{\delta}{2}\left[1-\bar{w}(\bar{x})^n\right] \qquad \text{for } \bar{w}(\bar{x})\geq 0 \tag{4.53}$$

$$\frac{d\bar{w}(\bar{x})}{d\bar{x}}=\frac{\delta}{2}\left[1+(2\sigma-1)(-\bar{w}(\bar{x}))^n\right] \qquad \text{for } \bar{w}(\bar{x})\leq 0 \tag{4.54}$$

This derivative is always positive due to the fact that $-1<\bar{w}(\bar{x})<1$ for all $-1\leq\bar{x}\leq 1$, $\sigma>1/2$ and $\delta>0$.

To see how Equations (4.53) and (4.54) have been obtained, define the function f as

$$f(\bar{x})=\varphi^+_{\sigma,n}\left[-\psi_{\sigma,n}(\delta)\right]+\frac{\delta}{2}(\bar{x}+1) \tag{4.55}$$

Then

$$\frac{\partial f(\bar{x})}{\partial \bar{x}} = \frac{\delta}{2} \tag{4.56}$$

Using the fact that the function $\psi^{+}_{\sigma,n}$ is the inverse of the function $\varphi^{+}_{\sigma,n}$ gives

$$\frac{\partial \psi^{+}_{\sigma,n}[f(\bar{x})]}{\partial \bar{x}} = \frac{\partial f(\bar{x})}{\partial \bar{x}} \left[\frac{\partial \psi^{+}_{\sigma,n}(\mu)}{\partial \mu} \right]_{\mu=f(\bar{x})} = \frac{\delta}{2} \frac{1}{\left[\frac{\partial \varphi^{+}_{\sigma,n}(v)}{\partial v} \right]_{v=\psi^{+}_{\sigma,n}[f(\bar{x})]}} \tag{4.57}$$

Using the fact that $\psi^{+}_{\sigma,n}[f(\bar{x})] = \bar{w}(x)$ (see Equation (4.5)), it follows from Equation (4.57) that

$$\frac{\partial \bar{w}(\bar{x})}{\partial \bar{x}} = \frac{\delta}{2} \frac{1}{\left[\frac{\partial \varphi^{+}_{\sigma,n}(v)}{\partial v} \right]_{v=\bar{w}(\bar{x})}} \tag{4.58}$$

Equations (4.53) and (4.54) follow from Equation (4.58) and Equations (3.20) and (3.23).

Figure 4.10 shows a typical curve $\bar{w}(\bar{x})$:

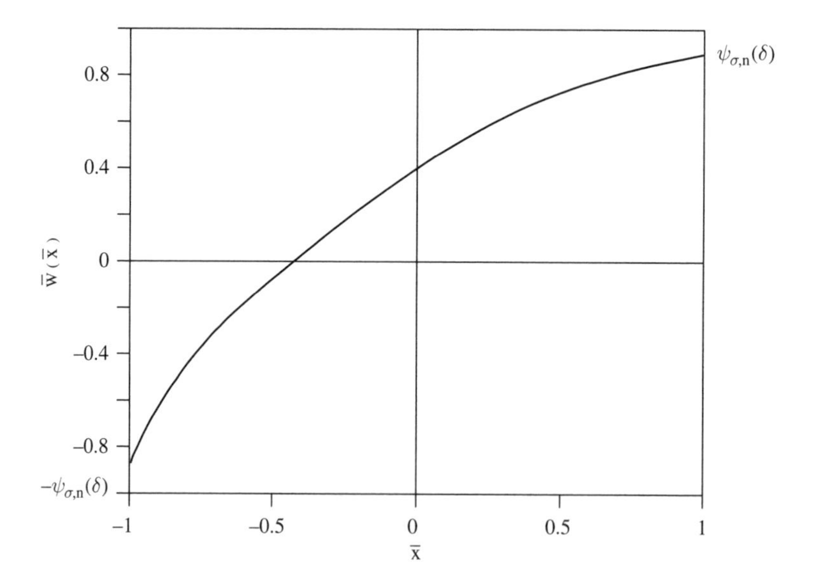

Figure 4.10 Variation of $\bar{w}(\bar{x})$ with the normalized input $\bar{x}$, with the values $\delta = 2$, $\sigma = 2$ and $n = 2$.

- The lowest value of the function $\bar{w}(\bar{x})$ is $-\psi_{\sigma,n}(\delta)$, which is obtained for $\bar{x} = -1$. This follows from Equation (4.5).
- The largest value of $\bar{w}(\bar{x})$ is $\psi_{\sigma,n}(\delta)$. It is obtained for $\bar{x} = 1$. This follows from Equation (4.12).

Since the function $\psi_{\sigma,n}(\delta)$ has been studied extensively in Section 4.3, the variation of $\bar{w}(\bar{x})$ for $\bar{x} = 1$ and $\bar{x} = -1$ will not be treated in this section. In other words, the fact that $-1 < \bar{x} < 1$ will be considered.

4.5.1 Variation with Respect to δ

Consider the function

$$\xi(\sigma, n, \delta, \bar{x}) = \varphi^{+}_{\sigma,n}\left[-\psi_{\sigma,n}(\delta)\right] + \frac{\delta}{2}(\bar{x}+1) \tag{4.59}$$

Note that $\xi(\sigma, n, \delta, \bar{x})$ is the argument of the function $\psi^{+}_{\sigma,n}(\cdot)$ in Equation (4.5). The following may be observed in Equation (4.59):

$$\xi(\sigma, n, 0, \bar{x}) = 0 \qquad \text{for } \delta = 0.$$
$$\lim_{\delta\to\infty} \xi(\sigma, n, \delta, \bar{x}) = \infty.$$

Using these two points in Equation (4.5) gives

$$\bar{w}(\bar{x}) = 0.$$
$$\lim_{\delta\to\infty} \bar{w}(\bar{x}) = 1.$$

On the other hand, deriving Equation (4.59) using (4.41) gives

$$\frac{\partial\xi(\sigma, n, \delta, \bar{x})}{\partial\delta} = \frac{\bar{x}+1}{2} - \frac{1}{1 + \dfrac{1+(2\sigma-1)\psi_{\sigma,n}(\delta)^n}{1-\psi_{\sigma,n}(\delta)^n}} \tag{4.60}$$

Now, two cases need to be discussed: (a) $0 \le \bar{x} < 1$ and (b) $-1 < \bar{x} < 0$.

The Case $0 \leq \bar{x} < 1$

In this case, it can be checked that

$$\frac{\partial \xi(\sigma, n, \delta, \bar{x})}{\partial \delta} > 0 \qquad \text{for all } \delta > 0$$

Since the function $\psi^{+}_{\sigma,n}(\cdot)$ is increasing with its argument, this means by Equation (4.5) that $\bar{w}(\bar{x})$ is increasing with δ. This behaviour is illustrated in Figure 4.11.

The case $-1 < \bar{x} < 0$

In this case, it can be checked that the hysteretic output $\bar{w}(\bar{x})$, seen as a function of the parameter δ, has a global minimum at δ^* given by

$$\delta^* = \varphi_{\sigma,n}\left[\sqrt[n]{\frac{-\bar{x}}{\sigma(\bar{x}+1)-\bar{x}}}\right] \tag{4.61}$$

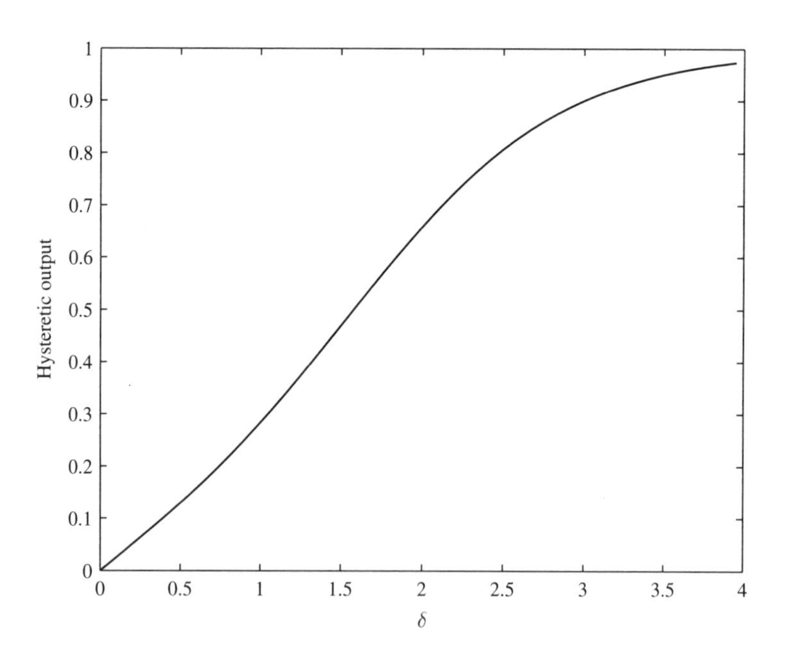

Figure 4.11 Variation of $\bar{w}(\bar{x})$ with δ for $\bar{x} = 0.5$, $\sigma = 1$ and $n = 2$.

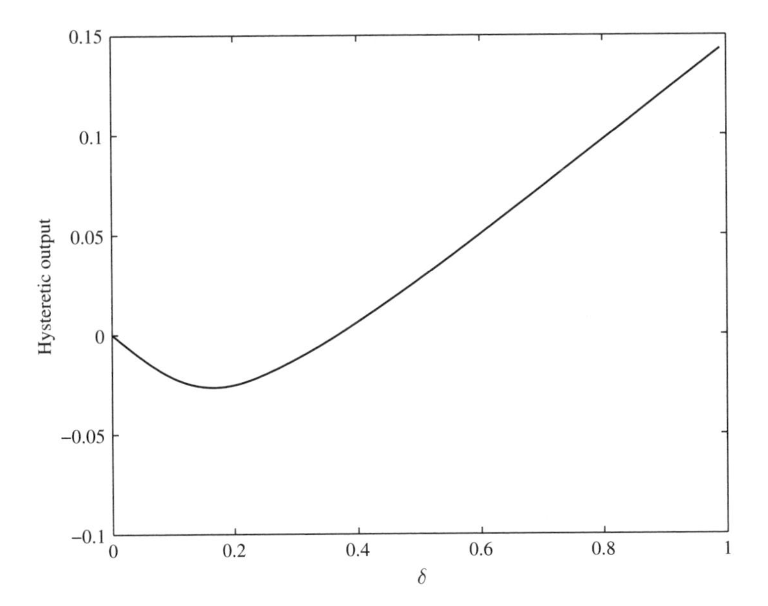

Figure 4.12 Variation of $\bar{w}(\bar{x})$ with δ for $\bar{x} = -0.5$, $\sigma = 100$ and $n = 2$.

Figure 4.12 gives in this case the evolution of the hysteretic output $\bar{w}(\bar{x})$ as a function of the parameter δ for $\bar{x} = -0.5$. It can be seen that the value $\delta^* = 0.1666$ corresponds to a global minimum.

4.5.2 Variation with Respect to σ

Given the fact that $\sigma \in [1/2, \infty)$, the behaviour of $\bar{w}(\bar{x})$ is first analyzed at the points $\sigma = 1/2$ and $\sigma = \infty$. At $\sigma = 1/2$, the value of the hysteretic output $\bar{w}(\bar{x})$ is obtained by putting $\sigma = 1/2$ in Equation (4.5). This value is denoted as

$$\bar{w}(\bar{x})_{\sigma=1/2}$$

To determine the asymptotic behaviour of $\bar{w}(\bar{x})$ with the parameter σ, note that, from Section 4.3.2,

$$\lim_{\sigma\to\infty} \psi_{\sigma,n}(\delta) = \psi_n^+(\delta)$$

The Lebesgue monotone convergence theorem [131, page 21] in Equation (3.23) shows that

$$\lim_{\sigma\to\infty} \varphi_{\sigma,n}^+[-\psi_n^+(\delta)] = 0$$

This implies by Equation (4.59) that, for σ sufficiently large, $\xi(\sigma, n, \delta, \bar{x}) > 0$, so that $\bar{w}(\bar{x}) = \psi^+_{\sigma,n}(\xi) = \psi^+_n(\xi)$ (see Equation (3.36)). As a conclusion, it is found that

$$\lim_{\sigma \to \infty} \bar{w}(\bar{x}) = \psi^+_n\left(\frac{\delta}{2}(\bar{x}+1)\right)$$

Now attention will be given to the variation of $\bar{w}(\bar{x})$ with the parameter σ. Using the composition rule in Equation (4.5) gives

$$\begin{aligned}\frac{\partial \bar{w}(\bar{x})}{\partial \sigma} = &\left[\frac{\partial \psi^+_{\sigma,n}(\mu)}{\partial \sigma}\right]_{\mu=\xi(\sigma,n,\delta,\bar{x})} + \left[\frac{\partial \psi^+_{\sigma,n}(\mu)}{\partial \mu}\right]_{\mu=\xi(\sigma,n,\delta,\bar{x})} \\ &\times \left\{\left[\frac{\partial \varphi^+_{\sigma,n}(w)}{\partial \sigma}\right]_{w=-\psi_{\sigma,n}(\delta)} - \left[\frac{\partial \varphi^+_{\sigma,n}(w)}{\partial w}\right]_{w=-\psi_{\sigma,n}(\delta)} \right. \\ &\left. \times \left[\frac{\partial \psi_{\sigma,n}(\mu)}{\partial \sigma}\right]_{\mu=\delta}\right\}\end{aligned} \tag{4.62}$$

In the following the different terms arising in Equation (4.62) will be computed:

$$\left[\frac{\partial \psi^+_{\sigma,n}(\mu)}{\partial \sigma}\right]_{\mu=\xi(\sigma,n,\delta,\bar{x})} = -\frac{\left[\frac{\partial \varphi^+_{\sigma,n}(w)}{\partial \sigma}\right]_{w=\bar{w}(\bar{x})}}{\left[\frac{\partial \varphi^+_{\sigma,n}(w)}{\partial w}\right]_{w=\bar{w}(\bar{x})}} \tag{4.63}$$

$$\left[\frac{\partial \psi^+_{\sigma,n}(\mu)}{\partial \mu}\right]_{\mu=\xi(\sigma,n,\delta,\bar{x})} = \frac{1}{\left[\frac{\partial \varphi^+_{\sigma,n}(w)}{\partial w}\right]_{w=\bar{w}(\bar{x})}} \tag{4.64}$$

where

$$\left[\frac{\partial \varphi^+_{\sigma,n}(w)}{\partial \sigma}\right]_{w=\bar{w}(\bar{x})} = 0 \qquad \text{for } \bar{w}(\bar{x}) \geq 0 \tag{4.65}$$

$$\left[\frac{\partial \varphi^+_{\sigma,n}(w)}{\partial \sigma}\right]_{w=\bar{w}(\bar{x})} = \int_0^{-\bar{w}(\bar{x})} \frac{2u^n}{[1+(2\sigma-1)u^n]^2}\,du \qquad \text{for } \bar{w}(\bar{x}) \leq 0 \tag{4.66}$$

$$\left[\frac{\partial \varphi^{+}_{\sigma,n}(w)}{\partial w}\right]_{w=\bar{w}(\bar{x})} = \frac{1}{1-\bar{w}(\bar{x})^n} \qquad \text{for } \bar{w}(\bar{x}) \geq 0 \tag{4.67}$$

$$\left[\frac{\partial \varphi^{+}_{\sigma,n}(w)}{\partial w}\right]_{w=\bar{w}(\bar{x})} = \frac{1}{1+(2\sigma-1)\,[-\bar{w}(\bar{x})]^n} \qquad \text{for } \bar{w}(\bar{x}) \leq 0 \tag{4.68}$$

$$\left[\frac{\partial \varphi^{-}_{\sigma,n}(w)}{\partial \sigma}\right]_{w=\bar{w}(\bar{x})} = -\int_0^{\bar{w}(\bar{x})} \frac{2u^n}{[1+(2\sigma-1)u^n]^2}\,\mathrm{d}u \qquad \text{for } \bar{w}(\bar{x}) \geq 0 \tag{4.69}$$

$$\left[\frac{\partial \varphi^{-}_{\sigma,n}(w)}{\partial \sigma}\right]_{w=\bar{w}(\bar{x})} = 0 \qquad \text{for } \bar{w}(\bar{x}) \leq 0 \tag{4.70}$$

$$\left[\frac{\partial \varphi^{-}_{\sigma,n}(w)}{\partial w}\right]_{w=\bar{w}(\bar{x})} = \frac{1}{1+(2\sigma-1)\bar{w}(\bar{x})^n} \qquad \text{for } \bar{w}(\bar{x}) \geq 0 \tag{4.71}$$

$$\left[\frac{\partial \varphi^{-}_{\sigma,n}(w)}{\partial w}\right]_{w=\bar{w}(\bar{x})} = \frac{1}{1-[-\bar{w}(\bar{x})]^n} \qquad \text{for } \bar{w}(\bar{x}) \leq 0 \tag{4.72}$$

Combining Equations (4.62) to (4.72) and (4.23) to (4.25) gives the following result for $\bar{w}(\bar{x}) \geq 0$:

$$\frac{\partial \bar{w}(\bar{x})}{\partial \sigma} = [1-\bar{w}(\bar{x})^n]\left(1 - \frac{1}{1+\dfrac{1+(2\sigma-1)\psi_{\sigma,n}(\delta)^n}{1-\psi_{\sigma,n}(\delta)^n}}\right) \times \int_0^{\psi_{\sigma,n}(\delta)} \frac{2u^n}{[1+(2\sigma-1)u^n]^2}\,\mathrm{d}u > 0 \tag{4.73}$$

This means that, for $\bar{w}(\bar{x}) \geq 0$, the term $\bar{w}(\bar{x})$ is increasing with σ. Note that the condition $\bar{w}(\bar{x}) \geq 0$ is equivalent to $\bar{x} \geq \bar{x}^{\circ}$. On the other hand, it was seen in Section 4.4.2 that $\bar{x}^{\circ}$ is a decreasing function of the parameter σ, which implies that

$$\bar{x}^{\circ} \text{ is a bijection from } \sigma \in [1/2, \infty) \text{ to } \left(-1, \frac{2}{\delta}\psi_{1/2,n}(\delta) - 1\right]$$

It follows then that two cases need to be discussed: (a) $\bar{x} \geq (2/\delta)\psi_{1/2,n}(\delta) - 1$ and (b) $\bar{x} < (2/\delta)\psi_{1/2,n}(\delta) - 1$.

The Case $\bar{x} \geq (2/\delta)\psi_{1/2,n}(\delta) - 1$

In this case, it follows from Equation (4.73) that

$$\frac{\partial \bar{w}(\bar{x})}{\partial \sigma} > 0 \qquad \text{for all } \bar{x} \geq \frac{2}{\delta}\psi_{1/2,n}(\delta) - 1 \text{ and for all } \sigma > 1/2$$

In other words, the function $\bar{w}(\bar{x})$ is increasing with the parameter σ (see Figure 4.13).

The Case $\bar{x} < (2/\delta)\psi_{1/2,n}(\delta) - 1$

It has been seen that the function

$$\bar{x}^{\circ} \text{is a bijection from } \sigma \in [1/2, \infty) \quad \text{to} \quad \left(-1, \frac{2}{\delta}\psi_{1/2,n}(\delta) - 1\right]$$

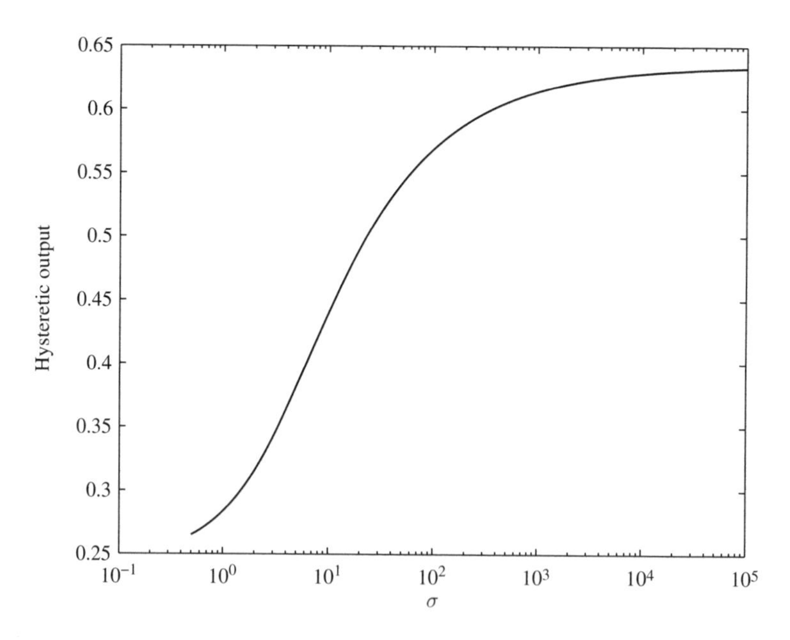

Figure 4.13 Variation of $\bar{w}(\bar{x})$ with σ for $\bar{x} = 0.5$, $\delta = 1$ and $n = 2$. In this case, $(2/\delta)\psi_{1/2,n}(\delta) - 1 = -0.0429$.

Using Equation (4.39), this implies that a unique value σ^{Δ} exists such that $\bar{x} = \bar{x}^{\circ}$, that is

$$\frac{2}{\delta}\varphi^{-}_{\sigma^{\Delta},n}\left[\psi_{\sigma^{\Delta},n}(\delta)\right] - 1 = \bar{x} \tag{4.74}$$

Given numerical values of $\bar{x}$, δ and n, the parameter σ^{Δ} can be found numerically using standard methods [132]. For $\sigma \geq \sigma^{\Delta}$, then $\bar{x}^{\circ} \leq \bar{x}$ as the function $\bar{x}^{\circ}$ is decreasing with σ. This means that $\bar{w}(\bar{x}) \geq 0$, so that, by Equation (4.73),

$$\frac{\partial \bar{w}(\bar{x})}{\partial \sigma} > 0$$

Thus, the function $\bar{w}(\bar{x})$ is increasing with σ in the interval $[\sigma^{\Delta}, \infty)$. The analysis of the variation of $\bar{w}(\bar{x})$ in the interval $[1/2, \sigma^{\Delta}]$ is difficult analytically, and for this reason numerical simulations are used to complete the picture. Figure 4.14 shows that a value $\bar{x}^{*}$ exists such that: for $\bar{x} \geq \bar{x}^{*}$ the function $\bar{w}(\bar{x})$ is increasing with the parameter σ and for $-1 < \bar{x} < \bar{x}^{*}$ the function $\bar{w}(\bar{x})$ has a global minimum.

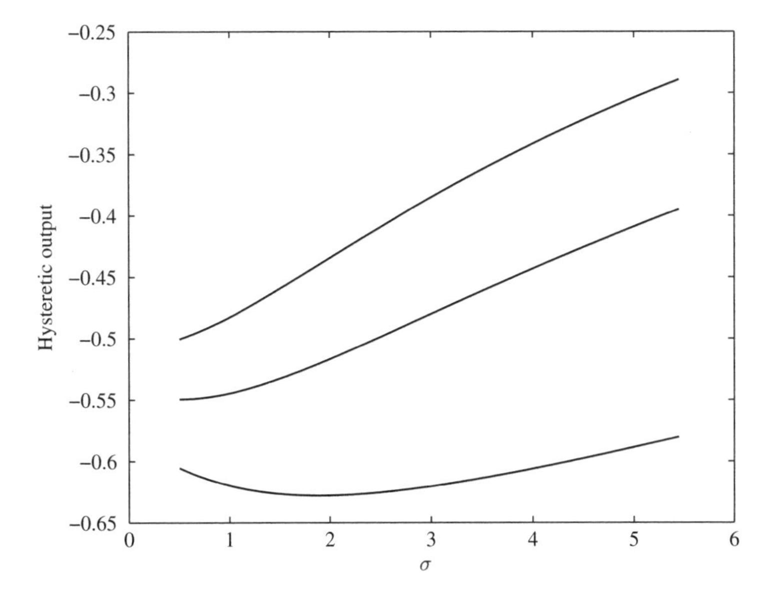

Figure 4.14 Variation of $\bar{w}(\bar{x})$ with σ for different values of $\bar{x}$, with $\delta = 1.4$ and $n = 2$. Upper curve, $\bar{x} = -0.8$; middle, $\bar{x} = -0.87$; lower, $\bar{x} = -0.95$. In this case. $(2/\delta)\psi_{1/2,n}(\delta) - 1 = -0.0851$ and $\bar{x}^{*} = -0.87$.

4.5.3 Variation with Respect to n

It is difficult to determine analytically the sign of

$$\frac{\partial \bar{w}(\bar{x})}{\partial n}$$

For this reason, the variation will be restricted to determining the limit of $\bar{w}(\bar{x})$ when $n \to \infty$. In this case, Lemma 8 shows that two cases need to be discussed: (a) $\delta \in (0, 2]$ and (b) $\delta > 2$.

The Case $\boldsymbol{\delta} \in (0, 2]$

In this case,

$$\lim_{n\to\infty} \bar{w}(\bar{x}) = \frac{\delta}{2}\,\bar{x}$$

This means that the hysteretic output $\bar{w}(\bar{x})$ approaches a linear behaviour asymptotically as the parameter n increases.

The Case $\boldsymbol{\delta} > 2$

In this case, the limit of $\bar{w}(\bar{x})$ is equal to the function defined by

$$\lim_{n\to\infty} \bar{w}(\bar{x}) = \begin{cases} -1 + \dfrac{\delta}{2}\,(\bar{x}+1) & \text{for } -1 \le \bar{x} < \dfrac{4}{\delta} - 1 \\ 1 & \text{for } \dfrac{4}{\delta} - 1 \le \bar{x} \le 1 \end{cases}$$

This function is represented in Figure 4.15 for the value $\delta = 5$. It is interesting to note that this function is independent of the parameter σ and depends only on the parameter δ.

4.5.4 Summary of the Obtained Results

This section summarizes the obtained results in Table 4.3. For example, the fourth row is to be read as follows:

Table 4.3 Variation of the hysteretic output with the Bouc–Wen model parameters δ, σ, n

$\bar{w}(\bar{x})$ with $\bar{x} \geq 0$	0	$\uparrow$		1
δ	0			$+\infty$
$\bar{w}(\bar{x})$ with $-1 < \bar{x} < 0$	0	$\downarrow$	$\bar{w}(\bar{x})_{\delta=\delta^*}$ $\uparrow$	1
δ	0		$\delta^* = \varphi_{\sigma,n}\left[\sqrt[n]{\dfrac{-\bar{x}}{\sigma(\bar{x}+1)-\bar{x}}}\right]$	$+\infty$
$\bar{w}(\bar{x})$ with $\bar{x} \geq \frac{2}{\delta}\psi_{1/2,n}(\delta) - 1$	$\bar{w}(\bar{x})_{\sigma=1/2}$	$\uparrow$		$\psi_n^+\left(\frac{\delta}{2}(\bar{x}+1)\right)$
σ	$\frac{1}{2}$			$+\infty$
$\bar{w}(\bar{x})$ with $\bar{x} < \frac{2}{\delta}\psi_{1/2,n}(\delta) - 1$	$\bar{w}(\bar{x})_{\sigma=1/2}$		$\bar{w}(\bar{x})_{\sigma=\sigma^\Delta}$ $\uparrow$	$\psi_n^+\left(\frac{\delta}{2}(\bar{x}+1)\right)$
σ	$\frac{1}{2}$		σ^Δ	$+\infty$
$\bar{w}(\bar{x})$	$\bar{w}(\bar{x})_{n=1}$			$\begin{cases} \frac{\delta}{2}\bar{x} & \text{if } \delta \in (0,2] \\ \begin{cases} -1+\frac{\delta}{2}(\bar{x}+1) & \text{for } -1 \leq \bar{x} < \frac{4}{\delta}-1 \\ 1 & \text{for } \frac{4}{\delta}-1 \leq \bar{x} \leq 1 \end{cases} & \text{if } \delta > 2 \end{cases}$
n	1			$+\infty$

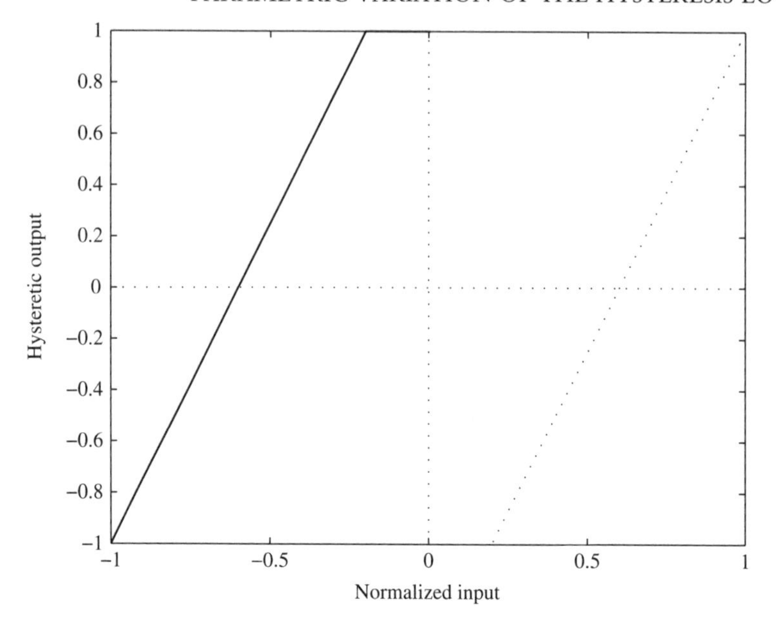

Figure 4.15 The limit function $\lim_{n\to\infty} \bar{w}(\bar{x})$ for $\delta = 5$.

- The first column says that the variation of $\bar{w}(\bar{x})$ is studied as a function of the parameter σ, in the case where $\bar{x} < (2/\delta)\psi_{1/2,n}(\delta) - 1$.
- The second column says that for $\sigma = 1/2$ the corresponding hysteretic output is $\bar{w}(\bar{x})_{\sigma=1/2}$ and for $\sigma = \sigma^{\Delta}$ the hysteretic output is $\bar{w}(\bar{x})_{\sigma=\sigma^{\Delta}}$. As stated in the previous section, it has not been possible to determine analytically the growth of the function $\bar{w}(\bar{x})$ (whether increasing $\uparrow$ or decreasing $\downarrow$) when $\sigma \in [1/2, \sigma^{\Delta}]$. For this reason, this growth is not reported in the table. On the contrary, the growth of $\bar{w}(\bar{x})$ has been determined analytically when $\sigma \in [\sigma^{\Delta}, \infty)$ and is reported in the table ($\uparrow$). Finally, the fourth line states that

$$\lim_{\sigma\to\infty} \bar{w}(\bar{x}) = \psi_n^+\left(\frac{\delta}{2}(\bar{x}+1)\right)$$

4.6 THE FOUR REGIONS OF THE BOUC–WEN MODEL

The linear and plastic behaviour of the hysteretic system is characterized by the derivative

$$\frac{d\bar{w}(\bar{x})}{d\bar{x}}$$

The value of this derivative is significant and almost constant in the linear part of the hysteretic limit cycle and is almost constant and much smaller in the plastic zone. In this section the notions of linear and plastic response in relation with the Bouc–Wen model are defined formally. The evolution of these regions are analysed using the model parameters.

4.6.1 The Linear Region R_l

The limit cycle of the hysteretic Bouc–Wen model is defined by Equations (4.9) to (4.12). It can also be seen as the unique solution of the differential equations (4.53) and (4.54) with the condition $\bar{w}(1) = \psi_{\sigma,n}(\delta)$. The right hand side of Equations (4.53) and (4.54) is composed of:

- A linear contribution $\delta/2$.
- A nonlinear contribution

$$\begin{cases} \frac{\delta}{2}(2\sigma - 1)\,[-\bar{w}(\bar{x})]^n & \text{for} \quad \bar{w}(\bar{x}) \leq 0 \\ -\frac{\delta}{2}\bar{w}(\bar{x})^n & \text{for} \quad \bar{w}(\bar{x}) \geq 0 \end{cases}$$

It is thus natural to consider that the linear behaviour of the hysteretic system corresponds to the linear contribution in Equations (4.53) and (4.54), while the nonlinear behaviour corresponds to the nonlinear contribution. Note that the nonlinear contribution becomes irrelevant for small values of the term $\bar{w}(\bar{x})$.

From Equations (4.53) and (4.54), it can be seen that, for these small values of $\bar{w}(\bar{x})$, the derivative of $\bar{w}(\bar{x})$ is close to

$$\left[\frac{d\bar{w}(\bar{x})}{d\bar{x}}\right]_{\bar{w}=0} = \frac{\delta}{2}$$

which implies that the linear region is characterized by the main slope $\delta/2$. The formal definition of the linear region in relation with the Bouc–Wen model is given below.

Definition 1. *Let $0 < r_1 < 1$ be a prescribed percentage (for example $r_1 = 0.01 = 1\,\%$). The linear region R_l is defined as the set of the points of the limit cycle such that the derivative at these points differs by no more than r_1 with respect to the main slope $\delta/2$. More precisely,*

$$R_l = \left\{ P = (\bar{x}, \bar{w}(\bar{x}))\ \textit{such that} \left| \frac{\left[\frac{d\bar{w}(\bar{x})}{d\bar{x}}\right]_P - \frac{\delta}{2}}{\frac{\delta}{2}} \right| \leq r_1 \right\} \tag{4.75}$$

The transition points $P_{sl} = (\bar{x}_{sl}, \bar{w}_{sl})$ and $P_{lt} = (\bar{x}_{lt}, \bar{w}_{lt})$ are defined as the points of the graph $(\bar{x}, \bar{w}(\bar{x}))$, where the value of the derivative of $\bar{w}(\bar{x})$ is different from its value at the point $\bar{w}(\bar{x}) = 0$ by r_1. The points P_{sl} and P_{lt} in Figure 4.3 correspond to the transitions from the region R_s to the region R_l and from the region R_l to the region R_t respectively. From Equations (4.53) and (4.54) it can be checked that these two points are defined by the following relations:

$$\bar{w}_{sl} = -n\sqrt{\frac{r_1}{2\sigma - 1}} \tag{4.76}$$

$$\bar{w}_{lt} = \sqrt[n]{r_1} \tag{4.77}$$

$$\bar{w}\,(\bar{x}_{sl}) = \bar{w}_{sl} \tag{4.78}$$

$$\bar{w}\,(\bar{x}_{lt}) = \bar{w}_{lt} \tag{4.79}$$

Analysing the influence of the Bouc–Wen normalized parameters on the shape of the hysteresis loop includes the analysis of the variation of the linear region with respect to these parameters. This analysis is equivalent to studing the evolution of the transition points $P_{sl} = (\bar{x}_{sl}, \bar{w}_{sl})$ and $P_{lt} = (\bar{x}_{lt}, \bar{w}_{lt})$ that define the linear region with respect to each parameter σ, δ and n. This means analysing the variation of the quantities $\bar{x}_{sl}$, $\bar{w}_{sl}$, $\bar{x}_{lt}$ and $\bar{w}_{lt}$.

The Different Possibilities for the Linear Region

Depending on the values of the parameters σ, n, δ and r_1, the hysteresis loop may present some peculiarities. Four cases are possible.

Case 1

The parameters are such that $\bar{w}_{lt} > \psi_{\sigma,n}(\delta)$. Note that the quantity $\psi_{\sigma,n}(\delta)$ corresponds to the largest amplitude of the hysteretic output, that is $\psi_{\sigma,n}(\delta) = \bar{w}(1)$. The term $\bar{w}_{lt}$ corresponds to the ordinate of the transition point P_{lt} from the linear region R_l to the region R_t (see Figure 4.3). If $\bar{w}_{lt} > \psi_{\sigma,n}(\delta)$, this means that the region of transition R_t and the plastic region are empty (see Figure 4.16). In this case, the linear region is defined on the axis of abscissas by the interval $[\bar{x}_{sl}, 1]$ and on the axis of ordinates by the interval $[\bar{w}_{sl}, \psi_{\sigma,n}(\delta)]$. The hysteresis loop presents at most two regions: the linear region R_l, and, possibly, the region of transition R_s. The condition $\bar{w}_{lt} > \psi_{\sigma,n}(\delta)$ can be written as

$$X_{\max} < \frac{\varphi_{\sigma,n}\left(\sqrt[n]{r_1}\right)}{2\rho} \triangleq X_0 \tag{4.80}$$

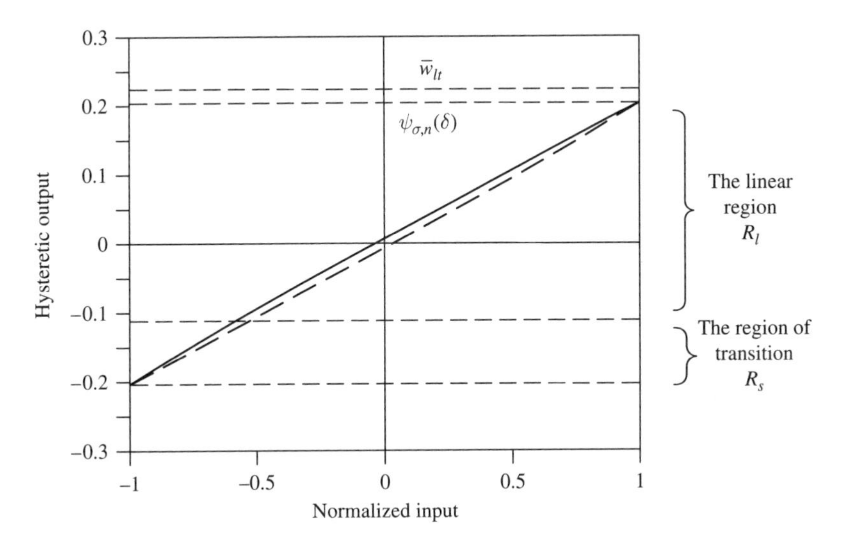

Figure 4.16 The linear region R_l in the case $\bar{w}_{lt} > \psi_{\sigma,n}(\delta)$.

Equation (4.80) shows that the specific shape of the hysteresis loop of Figure 4.16 is due to the small size of the displacement $x(t)$. This shape is obtained whenever $X_{\max} < X_0$, where the quantity X_0 depends on the Bouc–Wen model parameters σ, n, ρ and on the percentage r_1.

Case 2

The parameters are such that $\bar{w}_{sl} < -\psi_{\sigma,n}(\delta)$. Note that the absolute value of the quantity $-\psi_{\sigma,n}(\delta)$ corresponds to the largest amplitude of the hysteretic output, that is $\bar{w}(-1) = -\psi_{\sigma,n}(\delta)$. The term $\bar{w}_{sl}$ corresponds to the ordinate of the transition point P_{sl} from the region linear region R_l to the region R_s (see Figure 4.3). If $\bar{w}_{sl} < -\psi_{\sigma,n}(\delta)$, this means that the region of transition R_s is empty (see Figure 4.17). In this case, the linear region is defined on the axis of abscissas by the interval $[-1, \bar{x}_{lt}]$ and on the axis of ordinates by the interval $[-\psi_{\sigma,n}(\delta), \bar{w}_{lt}]$. The condition $\bar{w}_{sl} < -\psi_{\sigma,n}(\delta)$ can be written as

$$X_{\max} < \frac{\varphi_{\sigma,n}\left(\sqrt[n]{\frac{r_1}{2\sigma - 1}}\right)}{2\rho} \triangleq X_1 \tag{4.81}$$

Equation (4.81) shows that the specific shape of the hysteresis loop of Figure 4.17 is due to the small size of the displacement $x(t)$. This shape

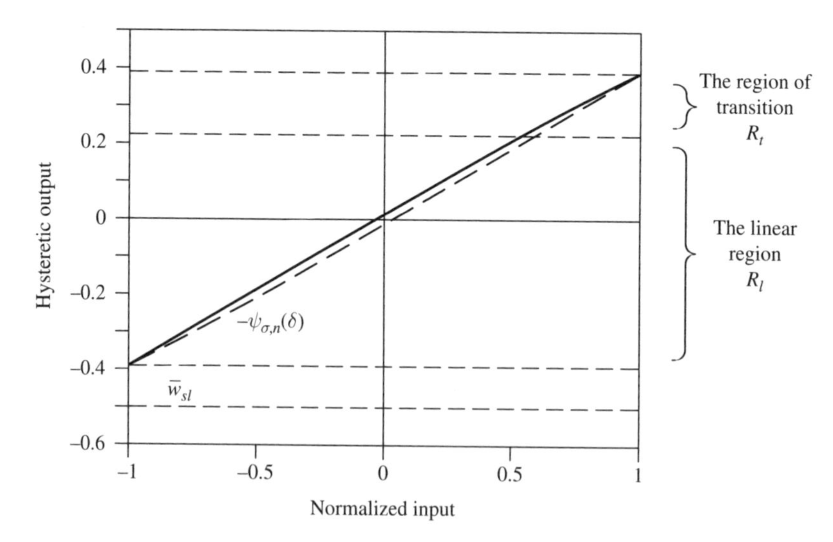

Figure 4.17 The linear region R_l in the case $\bar{w}_{sl} < -\psi_{\sigma,n}(\delta)$.

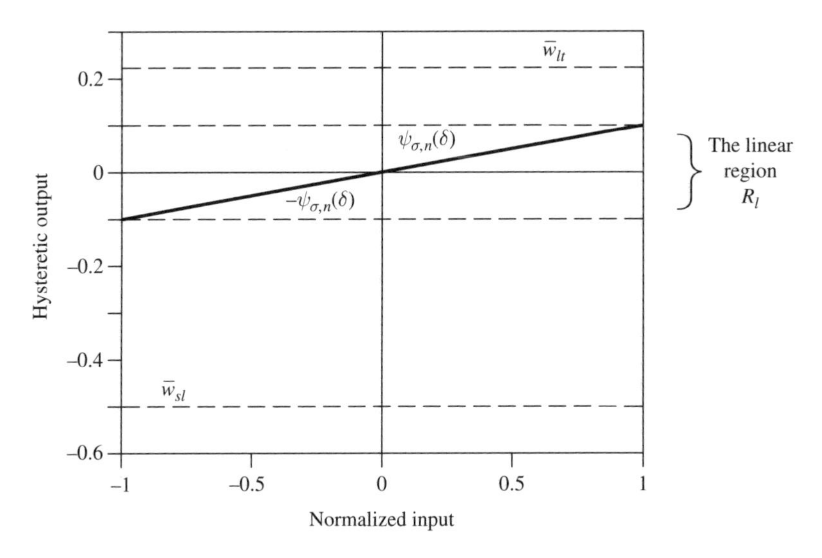

Figure 4.18 The linear region R_l in the case $\bar{w}_{sl} < -\psi_{\sigma,n}(\delta)$ and $\bar{w}_{lt} > \psi_{\sigma,n}(\delta)$.

is obtained whenever $X_{\max} < X_1$, where the quantity X_1 depends on the Bouc–Wen model parameters σ, n, ρ and on the percentage r_1.

Case 3

The parameters are such that $\bar{w}_{lt} > \psi_{\sigma,n}(\delta)$ and $\bar{w}_{sl} < -\psi_{\sigma,n}(\delta)$. This happens when cases 1 and 2 above occur simultaneously. In this case, the whole hysteresis loop reduces to the linear region (see Figure 4.18). The regions R_s, R_t and R_p are empty. In this case, the linear region is defined on the axis of abscissas by the interval $[-1, 1]$ and on the axis of ordinates by the interval $[-\psi_{\sigma,n}(\delta), \psi_{\sigma,n}(\delta)]$. This happens when

$$X_{\max} < \min(X_0, X_1) \triangleq X_2 \tag{4.82}$$

Note that the quantity X_2 depends on the Bouc–Wen model parameters σ, n, ρ and on the percentage r_1.

Case 4

The parameters are such that

$$-\psi_{\sigma,n}(\delta) \leq \bar{w}_{sl} < \bar{w}_{lt} \leq \psi_{\sigma,n}(\delta) \tag{4.83}$$

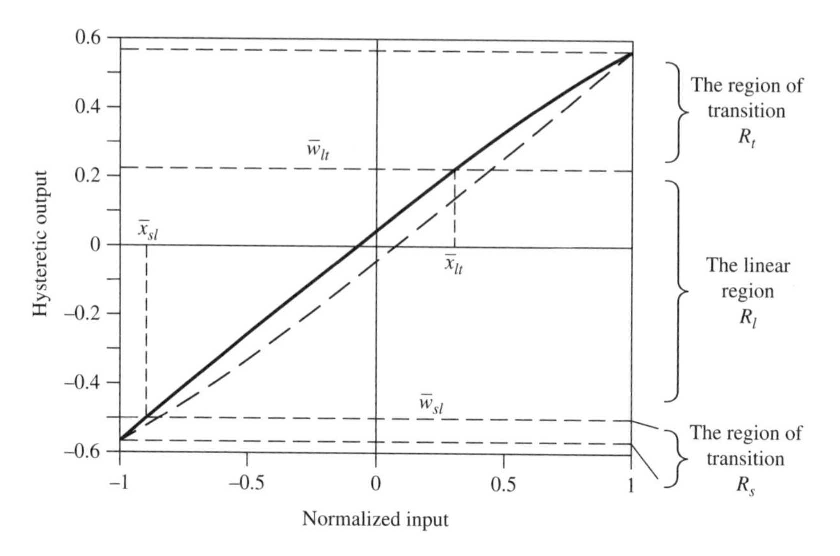

Figure 4.19 The linear region R_l in the case $-\psi_{\sigma,n}(\delta) \leq \bar{w}_{sl} < \bar{w}_{lt} \leq \psi_{\sigma,n}(\delta)$.

In this case, the hysteresis loop presents at least three nonempty regions as shown in Figure 4.19: the linear region R_l and two regions of transitions R_s and R_t (the plastic region may or may not be empty, as will be seen in Section 4.6.2). The linear region is then defined on the axis of abscissas by the interval $[\bar{x}_{sl}, \bar{x}_{lt}]$ and on the axis of ordinates by the interval $[\bar{w}_{sl}, \bar{w}_{lt}]$. This happens whenever

$$X_{\max} > \max(X_0, X_1) \triangleq X_3 \tag{4.84}$$

Implicit in Equation (4.84) is the assumption that both quantities X_0 and X_1 are finite. This is always the case for X_0; however, this is not always the case for X_1. An additional condition for the finiteness of X_1 is

$$\sigma > \frac{1+r_1}{2} \tag{4.85}$$

which comes as a condition for the argument of $\varphi_{\sigma,n}$ in Equation (4.81) to be less than unity. Inequalities (4.84) and (4.85) are equivalent to the inequalities in Equation (4.83) and insure that the regions R_s, R_l and R_t are nonempty. Note that the quantity X_3 in Equation (4.84) depends on the Bouc–Wen model parameters σ, n, ρ and on the percentage r_1. In the rest of the section, it will be assumed that both inequalities (4.84) and (4.85) hold.

Variation of the Region R_l with respect to δ

It can be seen from Equations (4.76) and (4.77) that the quantities $\bar{w}_{sl}$ and $\bar{w}_{lt}$ are independent of the parameter δ. This means that the linear region is independent of the parameter δ in the $\bar{w}$ axis. From Equations (4.5) and (4.76) to (4.79),

$$\bar{x}_{sl} = -1 + \frac{2}{\delta}\left\{\varphi^{+}_{\sigma,n}\left(-\sqrt[n]{\frac{r_1}{2\sigma-1}}\right) - \varphi^{+}_{\sigma,n}\left[-\psi_{\sigma,n}(\delta)\right\}\right\} \tag{4.86}$$

$$\bar{x}_{lt} = -1 + \frac{2}{\delta}\left\{\varphi^{+}_{\sigma,n}(\sqrt[n]{r_1}) - \varphi^{+}_{\sigma,n}\left[-\psi_{\sigma,n}(\delta)\right]\right\} \tag{4.87}$$

From Equations (4.86) and (4.83), it follows that $\bar{x}_{sl} \geq -1$. On the other hand, Equation (4.78) shows that $\bar{w}(\bar{x}_{sl}) < 0$. This fact, along with the increasing growth of $\bar{w}(\bar{x})$ with respect to $\bar{x}$, gives $\bar{x}_{sl} < \bar{x}^{\circ} < 0$, that is $1 \leq \bar{x}_{sl} < \bar{x}^{\circ} < 0$.

Now, from Equation (4.87) it follows that

$$\bar{x}_{lt} \leq -1 + \frac{2}{\delta}\left\{\varphi^{+}_{\sigma,n}\left[\psi_{\sigma,n}(\delta)\right] + \varphi^{-}_{\sigma,n}\left[\psi_{\sigma,n}(\delta)\right]\right\} \tag{4.88}$$

$$= -1 + \frac{2}{\delta}\left\{\varphi_{\sigma,n}\left[\psi_{\sigma,n}(\delta)\right]\right\} = 1 \tag{4.89}$$

Moreover, it follows from Equation (4.79) that $\bar{w}(\bar{x}_{lt}) > 0$, so that $\bar{x}_{lt} > \bar{x}^{\circ}$, that is $\bar{x}^{\circ} < \bar{x}_{lt} \leq 1$.

Note that

$$\lim_{\delta\to+\infty} \bar{x}_{sl} = \lim_{\delta\to+\infty} \bar{x}_{lt} = -1$$

which means that the size of the linear region along the axis of abscissas goes to zero as the parameter δ increases. As a conclusion, the points P_{sl} and P_{lt} that define the linear region go to the points

$$\left(-1, -\sqrt[n]{\frac{r_1}{2\sigma-1}}\right) \quad \text{and} \quad (-1, \sqrt[n]{r_1})$$

respectively, as $\delta \to \infty$.

Variation of the Region R_l with σ

Equations (4.76) and (4.86) show, respectively, that

$$\lim_{\sigma\to\infty} \bar{w}_{sl} = 0 \qquad \text{and} \qquad \lim_{\sigma\to\infty} \bar{x}_{sl} = -1$$

As σ increases, the part of the linear region R_l that corresponds to the negative ordinates shrinks and goes to the single point $(-1, 0)$. The quantity $\bar{w}_{lt}$ is independent of σ, so that the part of the linear region that corresponds to the positive ordinates remains constant. Equation (4.87) shows that

$$\lim_{\sigma\to\infty} \bar{x}_{lt} = -1 + \frac{2}{\delta}\, \varphi_n^+ \left(\sqrt[n]{r_1}\right)$$

As a conclusion, the points P_{sl} and P_{lt} that define the linear region go to the points

$$(-1, 0) \qquad \text{and} \qquad \left(-1 + \frac{2}{\delta}\varphi_n^+ \left(\sqrt[n]{r_1}\right), \sqrt[n]{r_1}\right)$$

respectively, as $\sigma \to \infty$.

Variation of the Region R_l with n

As seen in Table 4.3, when $n \to \infty$, the whole limit cycle becomes linear for $\delta \in (0, 2]$. For $\delta > 2$, the linear region R_l is exactly linear and corresponds to the interval

$$\left[-1, \frac{4}{\delta} - 1\right]$$

on the $\bar{x}$ axis and to the interval $[-1, 1]$ on the $\bar{w}$ axis.

As a conclusion, for $\delta \in (0, 2]$, the points P_{sl} and P_{lt} that define the linear region go to the points

$$\left(-1, -\frac{\delta}{2}\right) \qquad \text{and} \qquad \left(1, \frac{\delta}{2}\right)$$

respectively, as $n \to \infty$. For $\delta > 2$, the points P_{sl} and P_{lt} go to the points

$$(-1, -1) \qquad \text{and} \qquad \left(\frac{4}{\delta} - 1, 1\right)$$

respectively, as $n \to \infty$.

4.6.2 The Plastic Region R_p

The plastic region corresponds to the zone where large variations of the input induce small variations of the output; this is equivalent to saying that the derivative

$$\frac{\mathrm{d}\bar{w}(\bar{x})}{\mathrm{d}\bar{x}}$$

is small with respect to its value in the linear region R_l. The formal definition of the plastic region in relation to the Bouc–Wen model is given below.

Definition 2. *Let $0 < r_2 < 1$ be a prescribed percentage (for example $r_2 = 1\,\%$). The plastic region R_p is defined as the set of the points of the limit cycle such that the derivative at these points is no more than r_2 with respect to the main slope of the linear region. More precisely,*

$$R_p = \left\{ P = (\bar{x}, \bar{w}(\bar{x}))\; such\ that \left[\frac{\mathrm{d}\bar{w}(\bar{x})}{\mathrm{d}\bar{x}}\right]_P \le r_2 \frac{\delta}{2} \right\} \tag{4.90}$$

The plastic region is the part of the limit cycle that lies between the points $P_{tp} = (\bar{x}_{tp}, \bar{w}_{tp})$ and $P_p = (1, \psi_{\sigma,n}(\delta))$ (see Figure 4.3). The point P_{tp} defines the border between the region of transition R_t and the plastic region R_p. From Equation (4.53), the coordinates of the transition point P_{tp} are given as

$$\bar{w}_{tp} = \sqrt[n]{1 - r_2} \tag{4.91}$$

$$\bar{x}_{tp} = -1 + \frac{2}{\delta}\left\{\varphi^{+}_{\sigma,n}\left(\sqrt[n]{1 - r_2}\right) - \varphi^{+}_{\sigma,n}\left[-\psi_{\sigma,n}(\delta)\right]\right\} \tag{4.92}$$

For the plastic region to be nonempty, the ordinate $\psi_{\sigma,n}(\delta)$ of the point P_p should be greater than the ordinate $\bar{w}_{tp}$ of the point P_{tp}. This condition can be written as $\psi_{\sigma,n}(\delta) > \sqrt[n]{1-r_2}$ or equivalently

$$X_{\max} > \frac{\varphi_{\sigma,n}\left(\sqrt[n]{1-r_2}\right)}{2\rho} \triangleq X_4 \qquad (4.93)$$

Equation (4.93) means that, for the plastic region to be nonempty, the amplitude $X_{\max}$ of the input signal $x(t)$ should be larger than the quantity X_4. Note that X_4 depends on the Bouc–Wen model parameters σ, n, ρ and the prescribed percentage r_2.

In the following it is assumed that the inequality (4.93) holds. The evolution of the point P_p has been studied in Section 4.3, so that we concentrate solely on the analysis of the variation of the point P_{tp}.

Variation of the Region R_p with δ

The coordinate $\bar{w}_{tp}$ is independent of δ. This means that the plastic region is independent of the parameter δ in the $\bar{w}$ axis. On the other hand, it can be checked that

$$\bar{x}^{\circ} < \bar{x}_{tp} < 1 \qquad \text{and} \qquad \lim_{\delta\to\infty} \bar{x}_{tp} = -1$$

As a conclusion, the points P_{tp} and P_p that define the plastic region go to the points $\left(-1, \sqrt[n]{1-r_2}\right)$ and $(1, 1)$, respectively, as $\delta \to \infty$.

Variation of the Region R_p with σ

The coordinate $\bar{w}_{tp}$ is independent of σ. This means that the plastic region is independent of the parameter σ in the $\bar{w}$ axis. On the other hand, it can be checked that

$$\lim_{\sigma\to\infty} \bar{x}_{tp} = -1 + \frac{2}{\delta}\varphi_n^{+}\left(\sqrt[n]{1-r_2}\right)$$

As a conclusion, the points P_{tp} and P_p go to the points defined by the coordinates

$$\left(-1+\frac{2}{\delta}\varphi_n^+\left(\sqrt[n]{1-r_2}\right), \sqrt[n]{1-r_2}\right) \quad \text{and} \quad (1, \psi_n^+(\delta))$$

respectively, as $\sigma \to \infty$.

Variation of the region R_p with n

As seen in Table 4.3, when $n \to \infty$, the whole limit cycle becomes linear for $\delta \in (0, 2]$, so that the plastic region becomes empty. For $\delta > 2$, the points P_{tp} and P_p go to the points

$$\left(\frac{4}{\delta}-1, 1\right) \quad \text{and} \quad (1, 1)$$

respectively, as $n \to \infty$.

4.6.3 The Transition Regions R_t and R_s

The region R_s is defined by the points P_s and P_{sl} (see Figure 4.3). The point P_s is symmetric to P_p whose study has been done in Section 4.3. The variation of the point P_{sl} has been studied in Section 4.6.1. On the other hand, the transition region R_t is defined by the points P_{lt} and P_{tp} whose evolution has been studied in Sections 4.6.1 and 4.6.2 respectively.

4.7 INTERPRETATION OF THE NORMALIZED BOUC–WEN MODEL PARAMETERS

4.7.1 The Parameters ρ and δ

The shape of the limit cycle depends on the parameter $\delta = 2\rho X_{\max}$. This formula shows that the Bouc–Wen model parameter ρ and the maximal amplitude $X_{\max}$ of the periodic wave input signal have the same effect on the hysteresis loop. On the other hand, it was seen above that the slope of the linear region is $\delta/2 = \rho X_{\max}$. Thus, the parameter ρ can be interpreted as being the slope of the linear zone when the input has a unity maximal size, that is for $X_{\max} = 1$.

To derive an interpretation of the parameters from a mechanical point of view, note that Equation (2.5) leads to

$$\left(\frac{\mathrm{d}z}{\mathrm{d}x}\right)_{z=0} = \rho z_0 = A$$

in the unnormalized coordinates, as $\rho = A/z_0$ for $D = 1$. This means that A is the initial stiffness of the nonlinear component of the Bouc–Wen model, which is obtained by drawing the tangent to the z curve at the beginning of the deformation, as illustrated in Figure 4.20. This tangent crosses the horizontal line of the strength at a point whose projection on the horizontal axis is commonly called 'apparent yield point'. It can be checked that this value is equal to $1/\rho$. Thus, ρ can be interpreted as the inverse of the apparent yield point of the nonlinear component of the Bouc–Wen model.

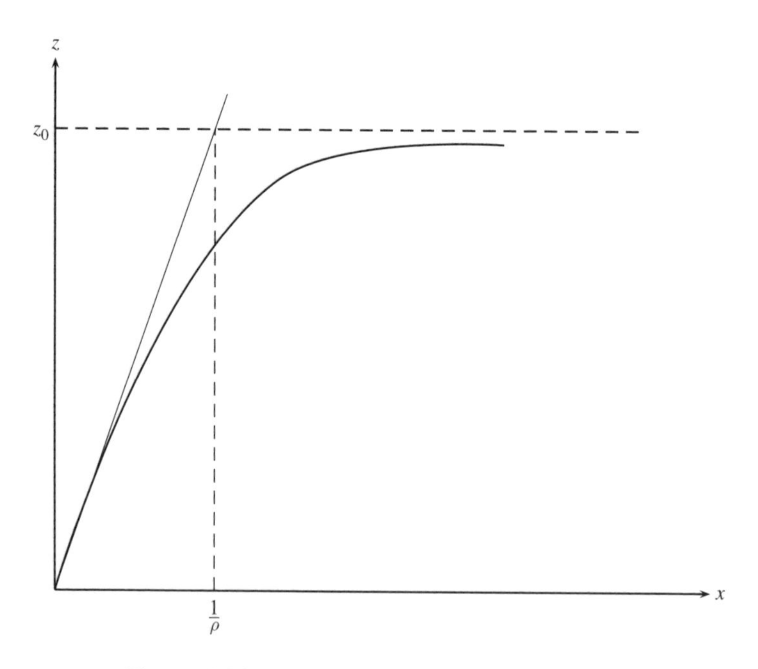

Figure 4.20 Interpretation of the parameter ρ.

To interpret the parameter δ, let us introduce $X_y = 1/\rho$, which means the *yield displacement*. Thus

$$\delta = 2\,\frac{X_{\max}}{X_y}$$

This indicates that the parameter δ measures the *ductility* of the model. This is a crucial variable in some applications in which a strong nonlinear response is expected, such as earthquake engineering.

4.7.2 The Parameter σ

Assume that the largest value $\psi_{\sigma,n}(\delta)$ of the hysteretic output is close to one. Then, for $\bar{x} = -1$ in Equation (4.54) it follows that

$$2\sigma = \frac{\left[\dfrac{\mathrm{d}\bar{w}(\bar{x})}{\mathrm{d}\bar{x}}\right]_{\bar{x}=-1}}{\delta/2} \tag{4.94}$$

Equation (4.94) says that the slope of the hysteresis loop for $\bar{x} = -1$ is 2σ times the slope of the linear region. This means that the parameter σ characterizes the slope of the limit cycle at the points $\bar{x} = -1$ and $\bar{x} = 1$ that correspond to changes in the sign of the velocity.

To derive an interpretation from the mechanical point of view, note that *softening* or *hardening* behaviours of the hysteretic material depend on whether the curvature of the graph $(\bar{x}, \bar{w}(\bar{x}))$ is concave of convex, respectively. To check this character the sign of the second derivative of $\bar{w}(\bar{x})$ should be studied. Equations (4.53) and (4.54) give

$$\frac{\mathrm{d}^2\bar{w}(\bar{x})}{\mathrm{d}\bar{x}^2} = -n\left(\frac{\delta}{2}\right)^2 [1 - \bar{w}(\bar{x})^n]\,\bar{w}(\bar{x})^{n-1} \qquad \text{for } \bar{w}(\bar{x}) \geq 0 \tag{4.95}$$

$$\frac{\mathrm{d}^2\bar{w}(\bar{x})}{\mathrm{d}\bar{x}^2} = -n(2\sigma - 1)\left(\frac{\delta}{2}\right)^2 \{1 + (2\sigma - 1)\,[-\bar{w}(\bar{x})]^n\}\,[-\bar{w}(\bar{x})]^{n-1} \qquad \text{for } \bar{w}(\bar{x}) \leq 0 \tag{4.96}$$

It can be seen from Equations (4.95) and (4.96) that the parameter σ is the one that shapes the curvature of the graph $(\bar{x}, \bar{w}(\bar{x}))$. Since $\sigma > 1/2$, this graph is concave (decreasing slope). This means that class I of the Bouc–Wen model corresponds to softening. As a consequence, it follows that, to have a Bouc–Wen model that describes the hardening behaviour, this model has to be either unstable or inconsistent with the laws of thermodynamics.

4.7.3 The Parameter n

This parameter has been investigated in various references by means of numerical simulations (see References [117] for example). A general conclusion that stems from these references is that the parameter n characterizes the transition from the linear to the plastic behaviour. In fact, this is only half true. Indeed, as seen above, the region of transition R_t from linear to plastic behaviour is defined by the points $P_{lt} = (\bar{x}_{lt}, \bar{w}_{lt})$ and $P_{tp} = (\bar{x}_{tp}, \bar{w}_{tp})$ (see also Figure 4.3). As seen in Equations (4.77) and (4.91), the ordinates $\bar{w}_{lt}$ and $\bar{w}_{tp}$ of the points P_{lt} and P_{tp} depend only on the parameter n (and on the prescribed percentages r_1 and r_2). This means that the transition region R_t is characterized only by the Bouc–Wen model parameter n along the axis of ordinates. However, Equations (4.87) and (4.92) show that the abscissas $\bar{x}_{lt}$ and $\bar{x}_{tp}$ of the points P_{lt} and P_{tp} depend not only on the parameter n but also on the rest of the parameters σ, ρ and also on the maximal size $X_{\max}$ of the input x.

As a conclusion, it follows that it is more precise to say that the parameter n is the one that characterizes the transition from linear to plastic behaviour along the axis of ordinates in the map $(\bar{x}, \bar{w}(\bar{x}))$.

4.8 CONCLUSION

This chapter has focused on analysing the manner in which the parameters of the normalized Bouc–Wen model influence the shape of the hysteresis loop. The study is based mainly on the analytical description of the limit cycle derived in Chapter 3.

The main practical results are summarized in three tables that describe the variation of some specific features of the hysteretic loop with the Bouc–Wen model parameters δ, σ, n:

- Table 4.1 gives the variation of the maximum value of the hysteretic output $\bar{w}(\bar{x})$.

- Table 4.2 describes the variation of the hysteretic zero $\bar{x}^{\circ}$, which is a measure of the hysteretic width $(2|\bar{x}_0|)$.
- Table 4.3 characterizes the variation of the hysteretic output $\bar{w}(\bar{x})$.

Additionally, four regions in the hysteretic loop have been identified, in particular those corresponding to linear and plastic behaviour respectively. The variation of these regions with respect to the model parameters has been analysed.

The analytical study of the hysteretic loop has also led to an interpretation of the parameters of the normalized Bouc–Wen model.

5

Robust Identification of the Bouc–Wen Model Parameters

5.1 INTRODUCTION

This chapter presents a parametric identification method for the normalized Bouc–Wen model and analyses its robustness to noise. Consider Figure 5.1 where the unknown vector θ is composed of the Bouc–Wen model parameters. The objective of any parameter identification algorithm is to determine an estimate $\hat{\theta}$ of the vector parameter θ using only the measurements of the input $x(t)$ and the output $\Phi_{\mathrm{BW}}(x)(t)$. It is assumed that the hysteretic state variable is not accessible to measurements as this state variable lacks a physical meaning. There are two types of parametric identification algorithms: recursive and nonrecursive. In a recursive algorithm, the estimate $\hat{\theta}$ is updated at each time instant t, while a nonrecursive algorithm gives the estimate $\hat{\theta}$ without the need for updating. The objective of a recursive algorithm is to ensure that $\lim_{t\to\infty} \hat{\theta}(t) = \theta$ in the absence of noise. The objective of a nonrecursive algorithm is to ensure that $\hat{\theta} = \theta$ in the absence of noise. In this case, it can be said that the identification algorithm possesses the estimation property.

In the current literature, much attention has been devoted to this problem. However, in all works that the authors are aware of, there is no proof that the estimated parameters converge (respectively, are equal) to the true ones in the case of recursive algorithms (respectively, nonrecursive algorithms); that is the estimation property is not

Systems with Hysteresis: Analysis, Identification and Control using the Bouc–Wen Model
F. Ikhouane and J. Rodellar

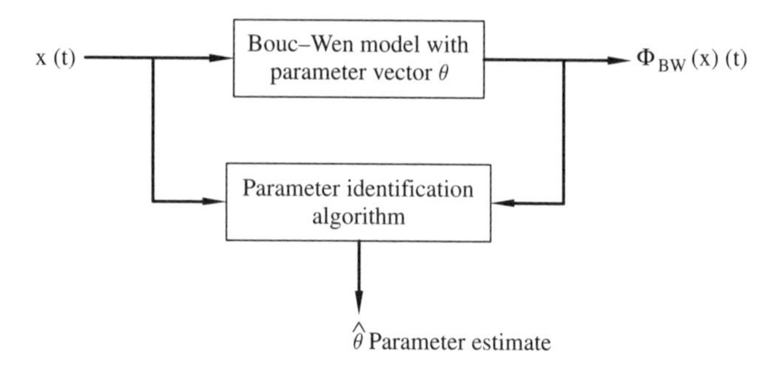

Figure 5.1 Parametric identification algorithm scheme.

established. This point is of crucial importance as the methodology that is used in these works is as follows:

1. Consider some existing identification technique that guarantees that the estimated parameters converge (or are equal) to the true ones, that is they possess the estimation property. This identification method is valid, in general, within a given context and under some assumptions.
2. Modify the identification method so that it fits with the problem of determining the Bouc–Wen model parameters. The modification of the method entails a modification of the context or the assumptions so that the estimation property is no longer guaranteed.
3. Instead of proving analytically the estimation property of the modified identification algorithm, the authors use a numerical simulation to check the validity of the estimation property.

While numerical simulations are useful in treating specific problems, they are useless for deriving general conclusions. More precisely, checking with a numerical simulation (or several simulations) the estimation property of an identification algorithm does not show that this algorithm does indeed possess that property. The only way to ensure that the estimation property is obtained for a given identification algorithm is to demonstrate it analytically.

In this chapter, a parametric nonlinear identification technique is proposed for the Bouc–Wen model based on the analytical description of Chapter 3. This method provides the exact values of the model parameters in the absence of disturbances and gives a guaranteed relative error between the estimated parameter and the true ones in the presence of perturbations. The identification technique consists in exciting the hysteretic system with two periodic signals that have a specific shape. The parameters of the Bouc–Wen model are then obtained from the two limit cycles using a precise algorithm. Moreover, it is shown that the identification technique is robust with respect to a class of disturbances of practical interest.

5.2 PARAMETER IDENTIFICATION OF THE BOUC–WEN MODEL

5.2.1 Class of Inputs

In this chapter, the input signal $x(t)$ is considered to be T-wave periodic, as presented in Section 3.2.1.

Assume that $|w(0)| \leq 1$. Then, by Table 3.1, $|w(t)| \leq 1$ for all $t \geq 0$. If a signal $x(t)$ is considered such that

$$\max\left(|X_{\max}|, |X_{\min}|\right) \gg \kappa_w/\kappa_x$$

then it follows from Equation (3.9) that the largest value

$$\kappa_x \max\left(|X_{\max}|, |X_{\min}|\right)$$

of the linear term $\kappa_x x(t)$ is much larger than the largest value κ_w of the nonlinear term $\kappa_w w(t)$. This means that the behaviour of Φ_{BW} versus x becomes almost linear for large values of the input x and that the hysteretic term $\kappa_w w(t)$ will have some influence on Φ_{BW} only for small values of the input $x(t)$. In particular, for large values of the input signal $x(t)$, the corresponding hysteretic output $\Phi_{\mathrm{BW}}(t)$ will be independent of the sign of the derivative $\dot{x}(t)$. This behaviour has not been reported experimentally for real hysteretic systems. For this reason, the Bouc–Wen model is not considered to represent a physical hysteretic behaviour if

$$\max\left(|X_{\max}|, |X_{\min}|\right) \gg \kappa_w/\kappa_x$$

Based on the above consideration, in this chapter only input signals such that

$$\max\left(|X_{\max}|, |X_{\min}|\right) \leq \kappa_w/\kappa_x$$

will be considered.

5.2.2 Identification Methodology

The hysteretic system under study is assumed to be described by the normalized Bouc–Wen model (3.9)–(3.10) with parameters $\kappa_x, \kappa_w, \rho, \sigma$ and n. The loading part of the limit cycle, which corresponds to an increasing input $x(t)$, can be obtained from Theorem 3 in the form

$$\bar{\Phi}_{\mathrm{BW}}(x) = \kappa_x x + \kappa_w \bar{w}(x) \tag{5.1}$$

$$\bar{w}(x) = \psi^{+}_{\sigma,n}\left(\varphi^{+}_{\sigma,n}\left[-\psi_{\sigma,n}\left(\rho\left(X_{\max} - X_{\min}\right)\right)\right] + \rho\left(x - X_{\min}\right)\right) \tag{5.2}$$

In general, the nonlinear state variable $\bar{w}(\cdot)$ is not accessible to measurement. However, in many cases of practical importance, the hysteretic limit cycle can be obtained experimentally. Here it is assumed that the hysteretic system is excited by the above class of periodic signals and the produced limit cycle is available. This means that the relation $\bar{\Phi}_{\mathrm{BW}}(x)$ resulting from Equations (5.1) and (5.2) is known. However, the parameters $\kappa_x, \kappa_w, \rho, \sigma$ and n in (5.1) and (5.2) are unknown. The objective of the proposed identification method is to determine the values of these parameters.

From Equation (5.2) it follows that (see Section 4.5 for a proof)

$$\frac{\mathrm{d}\bar{w}(x)}{\mathrm{d}x} = \rho\left(1 - \bar{w}(x)^n\right) \qquad \text{for } \bar{w}(x) \geq 0 \tag{5.3}$$

$$\frac{\mathrm{d}\bar{w}(x)}{\mathrm{d}x} = \rho\left(1 + (2\sigma - 1)\left[-\bar{w}(x)\right]^n\right) \qquad \text{for } \bar{w}(x) \leq 0 \tag{5.4}$$

Consider two wave T-periodic signals $x(t)$ and $x_1(t)$ such that $x_1(t) = x(t) + q$ for a given constant q. Denoting the corresponding hysteretic outputs $\bar{w}(x)$ and $\bar{w}_1(x_1)$, respectively, it follows from Equation (5.2) that $\bar{w}_1(x_1) = \bar{w}(x)$ for all $x \in [X_{\min}, X_{\max}]$. Then from Equation (5.1)

$$\bar{\Phi}_{\mathrm{BW},1}(x_1) = \bar{\Phi}_{\mathrm{BW}}(x) + \kappa_x q$$

This allows the value of κ_x to be determined in the form

$$\kappa_x = \frac{\bar{\Phi}_{\mathrm{BW},1}(x+q) - \bar{\Phi}_{\mathrm{BW}}(x)}{q} \tag{5.5}$$

for any value of $x \in [X_{\min}, X_{\max}]$.

Since κ_x has already been determined, the quantity $\kappa_w \bar{w}(x)$ can be computed from Equation (5.1) as

$$\kappa_w \bar{w}(x) = \bar{\Phi}_{\mathrm{BW}}(x) - \kappa_x x \triangleq \theta(x) \tag{5.6}$$

Knowledge of the function $\theta(x)$ for all $x \in [X_{\min}, X_{\max}]$ will allow the remaining parameters to be determined, as presented below.

From Equation (5.3) the following may be written:

$$\frac{\mathrm{d}\theta(x)}{\mathrm{d}x} = a - b\,\theta(x)^n \qquad \text{for } \theta(x) \geq 0 \tag{5.7}$$

where $a = \rho\kappa_w$ and $b = \rho\kappa_w^{-n+1}$. The coefficient a can be computed from Equation (5.7) as

$$a = \left[\frac{\mathrm{d}\theta(x)}{\mathrm{d}x}\right]_{x=x_*} \tag{5.8}$$

where x_* is such that $\theta(x_*) = 0$. The existence and uniqueness of this zero follows from the fact that the function $\bar{w}(x)$ is increasing from a negative value at $x = X_{\min}$ to a positive value at $x = X_{\max}$.

Now, take two design input values $x_{*2} > x_{*1} > x_*$. Evaluating the expression (5.7) for these two values, the parameter n and the quantity b can be determined as follows:

$$n = \frac{\log\left\{\dfrac{\left[\dfrac{\mathrm{d}\theta(x)}{\mathrm{d}x}\right]_{x=x_{*2}} - a}{\left[\dfrac{\mathrm{d}\theta(x)}{\mathrm{d}x}\right]_{x=x_{*1}} - a}\right\}}{\log\left[\dfrac{\theta(x_{*2})}{\theta(x_{*1})}\right]} \tag{5.9}$$

$$b = \frac{a - \left[\dfrac{\mathrm{d}\theta(x)}{dx}\right]_{x=x_{*2}}}{\theta(x_{*2})^n} \tag{5.10}$$

Further, the parameters κ_w and ρ are computed as follows:

$$\kappa_w = \sqrt[n]{\frac{a}{b}} \tag{5.11}$$

$$\rho = \frac{a}{\kappa_w} \tag{5.12}$$

Once the parameter κ_w has been determined, the function $\bar{w}(x)$ can be computed for all $x \in [X_{\min}, X_{\max}]$ from Equation (5.6) as

$$\bar{w}(x) = \frac{\theta(x)}{\kappa_w} \tag{5.13}$$

For the remaining parameter σ, Equation (5.4) may now be used. It can then be determined in the form

$$\sigma = \frac{1}{2}\left\{\frac{\dfrac{\left[\dfrac{\mathrm{d}\bar{w}(x)}{\mathrm{d}x}\right]_{x=x_{*3}}}{\rho} - 1}{[-\bar{w}(x_{*3})]^n} + 1\right\} \tag{5.14}$$

where x_{*3} is a design parameter such that $\bar{w}(x_{*3}) < 0$ or equivalently $x_{*3} < x_*$. This identification methodology can be systematically applied following the steps outlined in Table 5.1.

The identification procedure provides the exact values of the Bouc–Wen model parameters in the absence of disturbances. A by-product

Table 5.1 Procedure for identification of the Bouc–Wen model parameters

Step 1	Excite the Bouc–Wen model with a wave periodic signal $x(t)$. The output will reach a steady state $\bar{\Phi}_{BW}(t)$ as proved in Theorem 3. Since both input and output are measurable, the relation $(x, \bar{\Phi}_{BW}(x))$ is known.
Step 2	Choose a constant q and excite the Bouc–Wen model with the input $x_1(t) = x(t) + q$. The output will reach a steady state $\bar{\Phi}_{BW,1}(t)$. The relation $(x_1, \bar{\Phi}_{BW,1}(x_1))$ is known.
Step 3	Compute the coefficient κ_x using Equation (5.5).
Step 4	Compute the function $\theta(x)$ using Equation (5.6).
Step 5	Find the zero of $\theta(x)$, that is x_* such that $\theta(x_*) = 0$.
Step 6	Compute the parameter a using Equation (5.8).
Step 7	Choose constants x_{*1} and x_{*2} such that $x_{*2} > x_{*1} > x_*$. Compute parameters n and b using Equations (5.9) and (5.10).
Step 8	Compute parameters κ_w and ρ by Equations (5.11) and (5.12).
Step 9	Compute the function $\bar{w}(x)$ using Equation (5.13).
Step 10	Choose a constant x_{*3} such that $x_{*3} < x_*$. Compute parameter σ using Equation (5.14).

of this result is that, for the normalized form of the Bouc–Wen model, the relationship between the set of normalized parameters and the input/output behaviour is a bijection (see Section 3.3).

The next section analyses the robustness of the proposed identification method with respect to a class of disturbances of practical interest.

5.2.3 Robustness of the Identification Method

In practice, the T-periodic input signal $x(t)$ excites the hysteretic system by means of a (generally) linear actuator. Assume that the most significant frequency contents of the input signal lies within the bandwidth of this actuator and that the actuator has a static gain equal to one. Then the output of the actuator, after a transient, will reach a T-periodic steady state $x_d(t)$, which can be written as

$$x_d(t) = x(t) + d(t)$$

where the term can be understood as a perturbation produced by the actuator due to the fact that this actuator filters the high-frequency components of the input signal $x(t)$. Notice that $d(t)$ is T-periodic in the steady state. Moreover, since the static gain of the actuator is one, the maximum magnitude of $d(t)$ is small in comparison to the maximum value of the input signal $x(t)$.

On the other hand, T-periodic measurement disturbances result from the fact that a sensor has always a limited bandwidth. Thus, the high-frequency components of the hysteretic output are filtered, so that in the steady state, the measured output and the real output differ by a T-periodic function $v(t)$, as in the case of the input disturbances discussed in the previous paragraph.

Note that if the input disturbance $d(t)$ is not T-periodic (for example a random noise), then the limit cycle does not occur. Also, if the measurement disturbance $v(t)$ is not T-periodic, then even if the input disturbance $d(t)$ is T-periodic, the limit cycle is not observed. However, even though the necessity for the disturbances to be T-periodic constitutes the main theoretical limitation of the present identification method, experimental evidence shows that in many cases of practical relevance, limit cycles are indeed observed. This means that for these cases the most relevant disturbances are indeed T-periodic.

In this section it is considered that the input signal is corrupted by an additive disturbance which is constant or periodic with the same period as the input signal. More precisely, an unknown disturbance $d(t)$ is considered to be added to the input signal $x(t)$, resulting in a corrupted input signal

$$x_d(t) = x(t) + d(t)$$

If the signal $x_d(t)$ is accessible to measurement, then the analysis of the identification method is much easier as this is equivalent to identifying the Bouc–Wen model parameters with a known input and in the absence of disturbances. This case is included in the more general case of a signal $x(t)$ that is accessible to measurement and a signal $x_d(t)$ that is not accessible to measurement. This corresponds to an unknown signal $d(t)$, which is often the case in practice.

It is also considered that the hysteretic output $\bar{\Phi}_{BW}(t)$ is corrupted by an additive measurement disturbance $v(t)$ (see Figure 5.2). Note that it can be assumed without loss of generality that $X_{min} = -X_{max}$. The following assumption is made on the disturbances $d(t)$ and $v(t)$, where both are denoted by the generic notation $\xi(t)$.

Assumption 2. *The unknown disturbance signal $\xi(t)$ is constant or periodic of period T and is continuous for all $t \geq 0$ and C^1 on the interval $(0, T^+)\bigcup(T^+, T)$. Moreover, a constant $0 \leq \mu < 1/2$ exists such that*

$$|\xi(\tau)| \leq \mu X_{max} \qquad \textit{for } \tau \in [0, T] \tag{5.15}$$

$$\left|\dot{\xi}(\tau)\right| \leq \mu\,|\dot{x}(\tau)| \qquad \textit{for } \tau \in (0, T^+)\bigcup(T^+, T) \tag{5.16}$$

Clearly the disturbances d and v belong to the class of constant or small slowly time-varying periodic disturbances. These disturbances will be called μ-small. Note that, in practice, the perturbations may have high-frequency components that do not verify Assumption 2. In

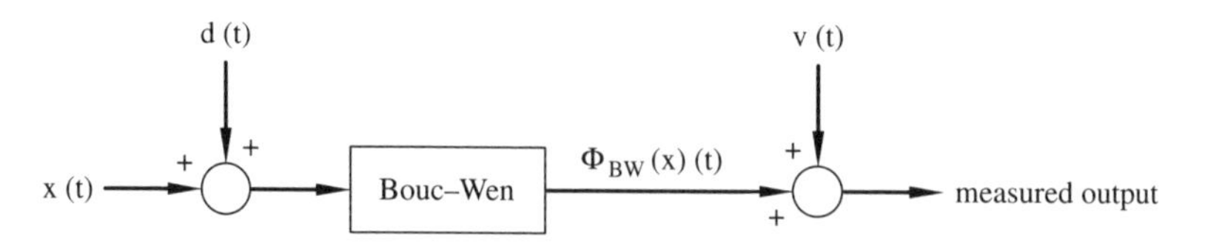

Figure 5.2 Identification in a noisy environment.

this case, a lowpass filter may be designed to eliminate these components so that the resulting perturbations comply with Assumption 2.

Under Assumption 2, the corrupted input signal $x_d(t)$ belongs to the class of inputs described in Section 5.2.1, so that limit cycles occur as described in Chapter 3 (Section 3.5). Denoting $X_{d,\max}$ and $X_{d,\min}$ the maximal and minimal values of x_d gives

$$X_{d,\max} = X_{\max} + d(T^+) \qquad \text{and} \qquad X_{d,\min} = -X_{\max} + d(0)$$

Then $\bar{x}(\tau)$ is considered as the normalized input function which is accessible to measurement:

$$-1 \leq \bar{x}(\tau) = \frac{x(\tau)}{X_{\max}} \leq 1$$

The (unknown) normalized disturbance are also considered:

$$\bar{\mathrm{d}}(\tau) = \frac{\mathrm{d}(\tau)}{X_{\max}} \qquad \text{and} \qquad \bar{v}(\tau) = \frac{v(\tau)}{X_{\max}}$$

Then define

$$\delta = 2\rho X_{\max} \qquad \text{and} \qquad \varepsilon_d = \frac{\bar{d}(T^+) - \bar{d}(0)}{2}$$

With these considerations, the limit cycle formulated by Theorem 3 (Section 3.5) can be written in the following form:

$$\begin{aligned} \bar{\Phi}_{\mathrm{BW}}(\tau) &= \kappa_x X_{\max}\bar{x}(\tau) + \kappa_x X_{\max}\bar{d}(\tau) \\ &\quad + \kappa_w \bar{w}(\tau) + X_{\max}\bar{v}(\tau) \qquad \text{for } \tau \in [0, T] \end{aligned} \tag{5.17}$$

$$\bar{w}(\tau) = \psi^{+}_{\sigma,n}\left(\varphi^{+}_{\sigma,n}\left[-\psi_{\sigma,n}\left(\delta\left(1+\varepsilon_d\right)\right)\right] + \frac{\delta}{2}\left[\bar{x}(\tau) + 1 + \bar{d}(\tau) - \bar{d}(0)\right]\right)$$
$$\text{for } \tau \in [0, T^+] \tag{5.18}$$

$$\bar{w}(\tau) = -\psi^{+}_{\sigma,n}\left(\varphi^{+}_{\sigma,n}\left[-\psi_{\sigma,n}\left(\delta\left(1+\varepsilon_d\right)\right)\right] - \frac{\delta}{2}\left[\bar{x}(\tau) - 1 + \bar{d}(\tau) - \bar{d}(T^+)\right]\right)$$
$$\text{for } \tau \in [T^+, T] \tag{5.19}$$

For any normalized Bouc–Wen model parameter p, the identified one is denoted p°. Let $0 < \varepsilon < 1/2$ be the maximal tolerance allowed for the identified parameters. This means that for each parameter p of the Bouc–Wen model and its corresponding identified value p°, then

$$\left|\frac{p - p^\circ}{p}\right| \leq \varepsilon$$

In this section, only the loading part of the hysteretic limit cycle is considered, that is in Equations (5.17) to (5.19) it is considered that $\tau \in [0, T^+]$. Since the input signal $\bar{x}(\tau)$ is, by assumption, such that $\dot{\bar{x}}(\tau) > 0$ for all $\tau \in (0, T^+)$, the function $\bar{x}(\tau)$ is a bijection from the time interval $[0, T^+]$ to $[-1, 1]$. Thus it is possible to define its inverse function from the interval $[-1, 1]$ to the time interval $[0, T^+]$. By an abuse of notation, this inverse function is denoted as τ.

With these notations, Equations (5.17) and (5.18) can be rewritten in the form

$$\bar{\Phi}_{\text{BW}}(\bar{x}) = \kappa_x X_{\max} \bar{x} + \kappa_x X_{\max} \bar{d}\,(\tau(\bar{x})) + \kappa_w \bar{w}(\bar{x}) + X_{\max} \bar{v}\,(\tau(\bar{x})) \quad (5.20)$$

$$\bar{w}(\bar{x}) = \psi^+_{\sigma,n}\left(\varphi^+_{\sigma,n}\left[-\psi_{\sigma,n}\left(\delta\left(1 + \varepsilon_d\right)\right)\right] + \frac{\delta}{2}\left[\bar{x} + 1 + \bar{d}\,(\tau(\bar{x})) - \bar{d}(0)\right]\right) \quad (5.21)$$

where the letter 'l' loading is not used for to simplify the notation. Also used are the notations

$$\bar{q} = \frac{q}{X_{\max}} \quad \text{and} \quad \bar{x}_{*i} = \frac{x_{*i}}{X_{\max}} \quad \text{for } i = 1, 2, 3$$

The main result of this section can now be stated.

Theorem 5. *Let $\bar{x}_{*1}$, $\bar{x}_{*2}$ and $\bar{x}_{*3}$ be design parameters and let $\varepsilon > 0$ be the desired precision on the estimated parameters. A real number $\mu^*\left(\kappa_x, \kappa_w, \rho, n, \sigma, \bar{q}, \bar{x}_{*1}, \bar{x}_{*2}, \bar{x}_{*3}, \varepsilon\right) > 0$ exists, called the robustness margin, such that*

(a) for any μ-small disturbances d and v verifying $0 \leq \mu \leq \mu^*$ and
(b) for any parameter $p \in \{\kappa_x, \kappa_w, \rho, n, \sigma\}$,

The corresponding identified parameter p° using the methodology of Section 5.2.2 is such that

$$\left|\frac{p-p^\circ}{p}\right| \leq \varepsilon$$

Robustness is a central issue in identification methods. It is not enough that the method gives the correct parameters in the absence of perturbations. It is also desirable that a 'small size' of disturbances leads to a 'small discrepancy' between the identified parameters and the true one. Theorem 5 states that, given a maximal tolerance $\varepsilon > 0$, for all μ-small disturbances such that $0 \leq \mu \leq \mu^*$, the relative error between the identified parameters p° and the true parameters p does not exceed ε. If the quantity μ^* were zero, this would imply that, even for arbitrarily small disturbances, the identification method could lead to a large discrepancy between the identified parameters and the true ones. Theorem 5 guarantees that the robustness margin is $\mu^* > 0$, so that all μ-small disturbances with $\mu \in [0, \mu^*]$ lead to a relative error in parameters no more than ε.

It is to be noted that, in practice, the (often unknown) size of the disturbance is a data of the problem, which means that μ is not a parameter chosen by a designer, but it is imposed by the physical process. This means that the identification method of Section 5.2.2 will guarantee that the relative error on the identified parameters is no larger than ε only for those values of ε large enough that the robustness margin μ^* is greater than the value of μ imposed by the physical process. Thus, the relative error on the parameters is, in practice, a result of the experimental identification and is determined a posteriori.

Proof. The proof is done in several steps:

1. Determination of the parameter κ_x.
2. Existence and unicity of the zero of the function $\theta^\circ(\bar{x})$.
3. Determination of the parameter n.
4. Determination of the parameter κ_w.
5. Determination of the parameter ρ.
6. Determination of the parameter σ.

Proof of step 1: Determination of the parameter κ_x. As seen in Section 5.2.2, determination of the parameter κ_x involves two experiments: the first one consists of obtaining the limit cycle with the input x and the second one consists of obtaining the limit cycle with the input $x_1 = x + q$. Both experiments are subject to input disturbances d and d_1 as well as to measurement disturbances v and v_1. All of these perturbations are supposed to satisfy Assumption 2. The equation of the limit cycle obtained with the input x_1 comes from Theorem 3 in the form

$$\begin{aligned}\bar{\Phi}_{\mathrm{BW},1}(x_1) = \kappa_x X_{\max}\bar{x} + \kappa_x X_{\max}\bar{d}_1(\tau(\bar{x})) \\ + \kappa_w \bar{w}_1(x_1) + X_{\max}\bar{v}_1(\tau(\bar{x})) + \kappa_x X_{\max}\bar{q}\end{aligned} \tag{5.22}$$

$$\begin{aligned}\bar{w}_1(x_1) = \psi^{+}_{\sigma,n}\left(\varphi^{+}_{\sigma,n}\left[-\psi_{\sigma,n}\left(\delta\left(1+\varepsilon_{1d}\right)\right)\right]\right. \\ \left. + \frac{\delta}{2}\left[\bar{x} + 1 + \bar{d}_1(\tau(\bar{x})) - \bar{d}_1(0)\right]\right)\end{aligned} \tag{5.23}$$

where

$$\varepsilon_{1d} = \frac{\bar{d}_1(T^{+}) - \bar{d}_1(0)}{2}, \qquad \bar{d}_1(\tau) = \frac{\mathrm{d}_1(\tau)}{X_{\max}} \qquad \text{and} \qquad \bar{v}_1(\tau) = \frac{v_1(\tau)}{X_{\max}}$$

To determine the relative error on the parameter κ_x, note that

$$\begin{aligned}&\left|\varphi^{+}_{\sigma,n}\left[-\psi_{\sigma,n}\left(\delta\left(1+\varepsilon_{1d}\right)\right)\right] - \varphi^{+}_{\sigma,n}\left[-\psi_{\sigma,n}\left(\delta\left(1+\varepsilon_{d}\right)\right)\right]\right| \\ &\quad = \left|\varepsilon_{1d} - \varepsilon_d\right| \left|\frac{\partial \varphi^{+}_{\sigma,n}\left[-\psi_{\sigma,n}\left(\delta\left(1+\varepsilon\right)\right)\right]}{\partial \varepsilon}\right|_{\varepsilon \in [\varepsilon_{1d}, \varepsilon_d]}\end{aligned} \tag{5.24}$$

On the other hand,

$$\left|\frac{\partial \varphi^{+}_{\sigma,n}\left[-\psi_{\sigma,n}\left(\delta\left(1+\varepsilon\right)\right)\right]}{\partial \varepsilon}\right|_{\varepsilon=\varepsilon_0 \in [\varepsilon_{1d}, \varepsilon_d]} = \left|\frac{\delta}{2\left[1 + \dfrac{\sigma \psi_{\sigma,n}\left(\delta\left(1+\varepsilon_0\right)\right)^n}{1 - \psi_{\sigma,n}\left(\delta\left(1+\varepsilon_0\right)\right)^n}\right]}\right| \tag{5.25}$$

By Assumption 2,

$$\left|\varepsilon_0\right| \le \max\left(\left|\varepsilon_{1d}\right|, \left|\varepsilon_d\right|\right) \le \mu < \frac{1}{2}$$

so that $\psi_{\sigma,n}(\delta(1+\varepsilon_0)) > 0$. This implies by Equation (5.25) that

$$\left|\frac{\partial\varphi^+_{\sigma,n}\left[-\psi_{\sigma,n}(\delta(1+\varepsilon))\right]}{\partial\varepsilon}\right|_{\varepsilon=\varepsilon_0\in[\varepsilon_{1d},\varepsilon_d]} \leq \frac{\delta}{2} \tag{5.26}$$

Combining Equations (5.24) and (5.26) along with Assumption 2 gives

$$\left|\varphi^+_{\sigma,n}\left[-\psi_{\sigma,n}(\delta(1+\varepsilon_{1d}))\right] - \varphi^+_{\sigma,n}\left[-\psi_{\sigma,n}(\delta(1+\varepsilon_d))\right]\right| \leq \mu\delta \tag{5.27}$$

On the other hand, from Equations (5.21) and (5.23) it is found that

$$|\bar{w}_1(x_1) - \bar{w}(\bar{x})| = |\zeta_0 - \zeta_1| \left|\frac{\partial\psi^+_{\sigma,n}(\zeta)}{\partial\zeta}\right|_{\zeta_2\in[\zeta_0,\zeta_1]} \tag{5.28}$$

where

$$\zeta_0 = \varphi^+_{\sigma,n}\left[-\psi_{\sigma,n}(\delta(1+\varepsilon_d))\right] + \frac{\delta}{2}\left[\bar{x} + 1 + \bar{d}(\tau(\bar{x})) - \bar{d}(0)\right] \tag{5.29}$$

$$\zeta_1 = \varphi^+_{\sigma,n}\left[-\psi_{\sigma,n}(\delta(1+\varepsilon_{1d}))\right] + \frac{\delta}{2}\left[\bar{x} + 1 + \bar{d}_1(\tau(\bar{x})) - \bar{d}_1(0)\right] \tag{5.30}$$

Combining Equations (5.27), (5.29) and (5.30) along with Assumption 2, it follows that

$$|\zeta_0 - \zeta_1| \leq 3\mu\delta \tag{5.31}$$

On the other hand, it follows from Equation (3.17) that

$$\left|\frac{\partial\psi^+_{\sigma,n}(\zeta)}{\partial\zeta}\right|_{\zeta_2} = 1 - \psi^+_{\sigma,n}(\zeta_2)^n \leq 1 \qquad \text{for } \zeta_2 \geq 0 \tag{5.32}$$

$$\left|\frac{\partial\psi^+_{\sigma,n}(\zeta)}{\partial\zeta}\right|_{\zeta_2} = 1 + (2\sigma - 1)\left[-\psi^+_{\sigma,n}(\zeta_2)\right]^n \leq 2\sigma \qquad \text{for } \zeta_2 < 0 \tag{5.33}$$

Using the fact that $\sigma \geq 1/2$, it follows from Equations (5.28) and (5.31) to (5.33) that

$$|\bar{w}_1(x_1) - \bar{w}(\bar{x})| \leq 6\mu\sigma\delta \tag{5.34}$$

The estimated parameter κ_x° is computed from Equation (5.5) as

$$\kappa_x^\circ = \frac{\bar{\Phi}_{BW,1}(x_1) - \bar{\Phi}_{BW}(\bar{x})}{q} \tag{5.35}$$

Combining Equations (5.35), (5.20) and (5.22), it follows that the relative error on the parameter κ_x is given by

$$\left|\frac{\kappa_x^\circ - \kappa_x}{\kappa_x}\right| \leq \frac{\left|\bar{d}_1 - \bar{d}\right|}{\bar{q}} + \frac{\kappa_w}{X_{\max}\kappa_x\bar{q}}\left|\bar{w}_1(x_1) - \bar{w}(\bar{x})\right| + \frac{|\bar{v}_1 - \bar{v}|}{\kappa_x\bar{q}} \tag{5.36}$$

Using Assumption 2, along with Equations (5.34) and (5.36), gives

$$\left|\frac{\kappa_x^\circ - \kappa_x}{\kappa_x}\right| \leq \frac{c_1\mu}{\bar{q}} \tag{5.37}$$

where

$$c_1 = 2 + \frac{2}{\kappa_x} + \frac{12\sigma\rho\kappa_w}{\kappa_x} \tag{5.38}$$

From inequality (5.37) it is clear that it is enough to have $\mu \leq \varepsilon\bar{q}/c_1$ to guarantee

$$\left|\frac{\kappa_x^\circ - \kappa_x}{\kappa_x}\right| \leq \varepsilon$$

The next step is to compute the function θ defined by Equation (5.6). However, the true value of the parameter κ_x is not known, but instead its estimate κ_x° is known. Thus, all that can be computed is the estimate

$$\theta^\circ(\bar{x}) = \bar{\Phi}_{BW}(\bar{x}) - \kappa_x^\circ X_{\max}\bar{x} \tag{5.39}$$

Proof of step 2: existence and unicity of the zero of the function $\theta^\circ(\bar{x})$. As seen in Section 5.2.2, determination of the rest of the parameters uses the zero of the function θ°. The existence of this zero will be ensured if it can be shown that $\theta^\circ(1) > 0$ and $\theta^\circ(-1) < 0$, due to the continuity of the function $\theta^{\circ}(\bar{x})$. From Equations (5.39), (5.20) and (5.19), it is found that

$$\theta^{\circ}(1) = (\kappa_x - \kappa_x^{\circ})\, X_{\max} + \kappa_x X_{\max} \bar{d}(T^+) + X_{\max} \bar{v}(T^+) + \kappa_w \psi_{\sigma,n}\left(\delta\left(1+\varepsilon_d\right)\right) \tag{5.40}$$

Using Assumption 2 along with Equations (5.40) and (5.37), it follows that

$$\theta^{\circ}(1) \geq -\mu X_{\max} c_2 + \kappa_w \psi_{\sigma,n}\left(\rho X_{\max}\right) \tag{5.41}$$

where

$$c_2 = \frac{c_1 \kappa_x}{\bar{q}} + \kappa_x + 1$$

The term $\psi_{\sigma,n}(\rho X_{\max})$ can be developed in a Taylor series as

$$\psi_{\sigma,n}\left(\rho X_{\max}\right) = \frac{1}{2} \rho X_{\max} + o\left(X_{\max}\right)$$

Thus, if

$$\mu < \frac{\rho \kappa_w}{2 c_2} = c_3$$

then, by Equation (5.41),

$$\theta^{\circ}(1) > 0 \qquad \text{for all } X_{\max} \in (0, A], \text{ where } A = f\left(\kappa_x, \kappa_w, \rho, \sigma, n, \bar{q}\right)$$

To have $\theta^{\circ}(1) > 0$ for all $X_{\max} \in (0, \kappa_w/\kappa_x]$, it is enough to have

$$\mu < \frac{\kappa_w \psi_{\sigma,n}\left(\rho A\right)}{c_2 (\kappa_w/\kappa_x)} = c_4$$

Similarly, it can be shown that

$$\theta^{\circ}(-1) < 0 \qquad \text{for all } X_{\max} \in \left(0, \frac{\kappa_w}{\kappa_x}\right]$$

for $\mu < c_5$, where

$$c_5 = f\left(\kappa_x, \kappa_w, \rho, \sigma, n, \bar{q}\right) > 0$$

Therefore, it has been proved that, for μ sufficiently small, the function $\theta^{\circ}(\bar{x})$ has at least one zero. It will now be shown that this zero is unique. To this end, the function $\bar{w}(\bar{x})$ is next proved to be strictly increasing. Indeed, define the map

$$g(\bar{x}) = \varphi^{+}_{\sigma,n}\left[-\psi_{\sigma,n}\left(\delta\left(1+\varepsilon_d\right)\right)\right] + \frac{\delta}{2}\left[\bar{x}+1+\bar{d}\left(\tau(\bar{x})\right)-\bar{d}(0)\right] \quad (5.42)$$

which appears in the right-hand side of Equation (5.21). Then, taking $\bar{x} \in (-1, 1)$ and differentiating with respect to $\bar{x}$, it follows that

$$\frac{\partial g(\bar{x})}{\partial \bar{x}} = \frac{\delta}{2}\left[1+\frac{\dot{\bar{d}}(\tau)}{\dot{\bar{x}}(\tau)}\right] \quad (5.43)$$

where $\dot{\bar{x}}$ and $\dot{\bar{d}}$ are set for the time derivatives and $\dot{\bar{x}}(\tau) \neq 0$ for all $\bar{x} \in (-1, 1)$ as the signal $\bar{x}$ is wave periodic. From Equation (5.43) and Assumption 2, it follows that

$$\frac{\partial g(\bar{x})}{\partial \bar{x}} \geq \frac{\delta}{4} > 0$$

This means that the function g is strictly increasing on the interval $[-1, 1]$. Since the function $\psi^{+}_{\sigma,n}$ is strictly increasing, this implies from Equation (5.21) that $\bar{w}(\bar{x})$ is a strictly increasing function of the argument $\bar{x}$.

On the other hand, the relationship between $\bar{w}(\bar{x})$ and $\bar{x}$ is given by Equation (5.21) and it can be checked that

$$\frac{\mathrm{d}\bar{w}(\bar{x})}{\mathrm{d}\bar{x}} = \frac{\delta}{2}\left[1+\frac{\dot{\bar{d}}\,(\tau)}{\dot{\bar{x}}\,(\tau)}\right]\left[1-\bar{w}(\bar{x})^{n}\right] \qquad \text{for } \bar{w}(\bar{x}) \geq 0 \quad (5.44)$$

$$\frac{\mathrm{d}\bar{w}(\bar{x})}{\mathrm{d}\bar{x}} = \frac{\delta}{2}\left[1+\frac{\dot{\bar{d}}\,(\tau)}{\dot{\bar{x}}\,(\tau)}\right]\left\{1+(2\sigma-1)\left[-\bar{w}(\bar{x})\right]^{n}\right\} \qquad \text{for } \bar{w}(\bar{x}) \leq 0 \quad (5.45)$$

whenever the time derivative $\dot{\bar{x}}(\tau)$ is nonzero. Note that since the signal $\bar{x}$ is wave periodic, the derivative $\dot{\bar{x}}(\tau)$ may be zero only at the points $\tau = 0$ and $\tau = T^{+}$, which correspond to the values $\bar{x} = -1$ and $\bar{x} = 1$, respectively.

Combining Equations (5.39), (5.20), (5.44) and (5.45), it follows that

$$\frac{\partial\theta^\circ(\bar{x})}{\partial\bar{x}} = \kappa_x X_{\max}\left[\frac{\kappa_x-\kappa_x^\circ}{\kappa_x}+\frac{\dot{\bar{d}}(\tau)}{\dot{\bar{x}}(\tau)}+\frac{\dot{\bar{v}}(\tau)}{\kappa_x\dot{\bar{x}}(\tau)}\right] + \rho\kappa_w X_{\max}\left[1+\frac{\dot{\bar{d}}(\tau)}{\dot{\bar{x}}(\tau)}\right][1-\bar{w}(\bar{x})^n] \quad \text{for } \bar{w}(\bar{x})\geq 0 \tag{5.46}$$

$$\frac{\partial\theta^\circ(\bar{x})}{\partial\bar{x}} = \kappa_x X_{\max}\left[\frac{\kappa_x-\kappa_x^\circ}{\kappa_x}+\frac{\dot{\bar{d}}(\tau)}{\dot{\bar{x}}(\tau)}+\frac{\dot{\bar{v}}(\tau)}{\kappa_x\dot{\bar{x}}(\tau)}\right] + \rho\kappa_w X_{\max}\left[1+\frac{\dot{\bar{d}}(\tau)}{\dot{\bar{x}}(\tau)}\right]\{1+(2\sigma-1)[-\bar{w}(\bar{x})]^n\} \quad \text{for } \bar{w}(\bar{x})\leq 0 \tag{5.47}$$

Using the fact that $\bar{w}(1)=\psi_{\sigma,n}(\delta(1+\varepsilon_d))$ and that $X_{\max}\leq\kappa_w/\kappa_x$ (see Section 5.2.1), along with Assumption 2, it follows from Equations (5.46) and (5.47) that

$$\frac{\partial\theta^\circ(\bar{x})}{\partial\bar{x}} \geq X_{\max}(-\mu c_2+c_6) \tag{5.48}$$

where

$$c_6=\frac{\rho\kappa_w}{2}\left[1-\psi_{\sigma,n}\left(\frac{3\rho\kappa_w}{\kappa_x}\right)^n\right]>0 \tag{5.49}$$

Taking $\mu\leq c_6/(2c_2)$, the following is obtained from Equation (5.48):

$$\frac{\partial\theta^\circ(\bar{x})}{\partial\bar{x}} \geq \frac{c_6}{2}=c_7>0 \tag{5.50}$$

From Equation (5.50), it is found that the function $\theta^\circ(\bar{x})$ is strictly increasing, which proves the unicity of its zero, which is denoted $\bar{x}_*^\circ$.

Proof of step 3: determination of the parameter n. It has been proved above that the nonlinear function $\theta^\circ(\bar{x})$ is strictly increasing from a negative value at $\bar{x}=-1$ to a positive value at $\bar{x}=1$. Since this function is computable, standard numerical methods can determine its zero $\bar{x}_*^\circ$ with the (in general good) precision of the computer. It is to be noted that, by combining Equations (5.39) and (5.37), together with the fact that $\theta^\circ(\bar{x}_*^\circ) = 0$, then

$$|\bar{w}(\bar{x}_*^\circ)| \leq \frac{c_2}{\kappa_x}\mu = c_8\mu \tag{5.51}$$

Now, let

$$\bar{x}_{*1} = \bar{x}_*^\circ + r_1(1 - \bar{x}_*^\circ) \qquad \text{where } 0 < r_1 < 1$$

is a design parameter. Then, from Equation (5.50), it is found that

$$\theta^\circ(\bar{x}_{*1}) \geq c_7(\bar{x}_{*1} - \bar{x}_*^\circ) = c_7 r_1(1 - \bar{x}_*^\circ) = c_9 \tag{5.52}$$

Due to the fact that the function θ° is increasing, then $\theta^\circ(\bar{x}) \geq c_9$ for all $\bar{x} \geq \bar{x}_{*1}$. This fact along with Equations (5.39) and (5.37) and Assumption 2, gives

$$\bar{w}(\bar{x}) > 0 \quad \text{for all } \bar{x} \geq \bar{x}_{*1} \text{ and } \mu \leq \frac{\kappa_x c_9}{2\kappa_w c_2}$$

Let $\bar{x}_{*2} = \bar{x}_*^\circ + r_2(1 - \bar{x}_*^\circ)$ where $0 < r_1 < r_2 < 1$ is a design parameter. An estimate of the parameter n comes from Equation (5.9) as follows:

$$n^\circ = \frac{\log\left\{\dfrac{\left[\dfrac{d\theta^\circ(\bar{x})}{d\bar{x}}\right]_{\bar{x}=\bar{x}_{*2}} - \left[\dfrac{d\theta^\circ(\bar{x})}{d\bar{x}}\right]_{\bar{x}=\bar{x}_*^\circ}}{\left[\dfrac{d\theta^\circ(\bar{x})}{d\bar{x}}\right]_{\bar{x}=\bar{x}_{*1}} - \left[\dfrac{d\theta^\circ(\bar{x})}{d\bar{x}}\right]_{\bar{x}=\bar{x}_*^\circ}}\right\}}{\log\left[\dfrac{\theta^\circ(\bar{x}_{*2})}{\theta^\circ(\bar{x}_{*1})}\right]} \tag{5.53}$$

Define the function

$$\begin{aligned}\vartheta_1(\bar{x}) = {} & \frac{\kappa_x - \kappa_x^\circ}{\kappa_x} + \frac{\dot{\bar{d}}(\tau(\bar{x}))}{\dot{\bar{x}}(\tau(\bar{x}))} + \frac{\dot{\bar{v}}(\tau(\bar{x}))}{\kappa_x \dot{\bar{x}}(\tau(\bar{x}))} \\ & + \frac{\rho\kappa_w}{\kappa_x} \times \frac{\dot{\bar{d}}(\tau(\bar{x}))}{\dot{\bar{x}}(\tau(\bar{x}))}[1 - \bar{w}(\bar{x})^n] \qquad \text{for } \bar{w}(\bar{x}) \geq 0\end{aligned}$$

$$\vartheta_1(\bar{x}) = \frac{\kappa_x - \kappa_x^\circ}{\kappa_x} + \frac{\dot{\bar{d}}(\tau(\bar{x}))}{\dot{\bar{x}}(\tau(\bar{x}))} + \frac{\dot{\bar{v}}(\tau(\bar{x}))}{\kappa_x \dot{\bar{x}}(\tau(\bar{x}))} + \frac{\rho\kappa_w}{\kappa_x} \times \frac{\dot{\bar{d}}(\tau(\bar{x}))}{\dot{\bar{x}}(\tau(\bar{x}))} (1 + (2\sigma - 1)(-\bar{w}(\bar{x}))^n) \qquad \text{for } \bar{w}(\bar{x}) \leq 0 \tag{5.54}$$

Then, it can be checked that

$$|\vartheta_1(\bar{x})| \leq \frac{1}{\kappa_x} (c_2 + 2\rho\kappa_w\sigma)\,\mu = c_{10}\mu \tag{5.55}$$

On the other hand, the following is obtained for $\bar{w}(\bar{x}_*^\circ) \geq 0$:

$$\frac{\left[\frac{\mathrm{d}\theta^\circ(\bar{x})}{\mathrm{d}\bar{x}}\right]_{\bar{x}=\bar{x}_{*2}} - \left[\frac{\mathrm{d}\theta^\circ(\bar{x})}{\mathrm{d}\bar{x}}\right]_{\bar{x}=\bar{x}_*^\circ}}{\left[\frac{\mathrm{d}\theta^\circ(\bar{x})}{\mathrm{d}\bar{x}}\right]_{\bar{x}=\bar{x}_{*1}} - \left[\frac{\mathrm{d}\theta^\circ(\bar{x})}{\mathrm{d}\bar{x}}\right]_{\bar{x}=\bar{x}_*^\circ}} = \frac{\vartheta_1(\bar{x}_{*2}) - \vartheta_1(\bar{x}_*^\circ) + \frac{\rho\kappa_w}{\kappa_x}[\bar{w}(\bar{x}_*^\circ)^n - \bar{w}(\bar{x}_{*2})^n)}{\vartheta_1(\bar{x}_{*1}) - \vartheta_1(\bar{x}_*^\circ) + \frac{\rho\kappa_w}{\kappa_x}(\bar{w}(\bar{x}_*^\circ)^n - \bar{w}(\bar{x}_{*1})^n]}$$

$$= \frac{\bar{w}(\bar{x}_{*2})^n}{\bar{w}(\bar{x}_{*1})^n}\left(\frac{1+\mu f_1}{1+\mu f_2}\right) = \frac{\bar{w}(\bar{x}_{*2})^n}{\bar{w}(\bar{x}_{*1})^n}(1+\mu f_3) \tag{5.56}$$

where

$$c_{12} = \frac{c_{11}}{\bar{w}(\bar{x}_{*2})^n} + 2\frac{c_{11}}{\bar{w}(\bar{x}_{*1})^n} + \frac{c_{11}^2}{[\bar{w}(\bar{x}_{*1})\bar{w}(\bar{x}_{*2})]^n} \tag{5.57}$$

$$c_{11} = \frac{2c_{10}\kappa_x}{\rho\kappa_w} + 2\sigma c_8^n \tag{5.58}$$

$$|f_1| \leq \frac{c_{11}}{\bar{w}(\bar{x}_{*2})^n} \tag{5.59}$$

$$|f_2| \leq \frac{c_{11}}{\bar{w}(\bar{x}_{*1})^n} \tag{5.60}$$

$$|f_3| \leq c_{12} \tag{5.61}$$

Similar relations are obtained for $\bar{w}(\bar{x}_*^\circ) \leq 0$. Define the function

$$\vartheta_2(\bar{x}) = \kappa_x X_{\max}\left[\frac{\kappa_x - \kappa_x^\circ}{\kappa_x} + \frac{\dot{\bar{d}}(\tau(\bar{x}))}{\dot{\bar{x}}(\tau(\bar{x}))} + \frac{\dot{\bar{v}}(\tau(\bar{x}))}{\kappa_x\dot{\bar{x}}(\tau(\bar{x}))}\right] \tag{5.62}$$

Then $|\vartheta_2(\bar{x})| \leq c_{13}\mu$, where $c_{13} = \kappa_w c_2/\kappa_x$. With these notations the following equation is obtained in a similar way:

$$\frac{\theta^\circ(\bar{x}_{*2})}{\theta^\circ(\bar{x}_{*1})} = \frac{\vartheta_2(\bar{x}_{*2}) + \kappa_w \bar{w}(\bar{x}_{*2})}{\vartheta_2(\bar{x}_{*1}) + \kappa_w \bar{w}(\bar{x}_{*1})} = \frac{\bar{w}(\bar{x}_{*2})}{\bar{w}(\bar{x}_{*1})}(1 + \mu f_4) \tag{5.63}$$

where

$$|f_4| \leq 3c_8 + c_8^2 = c_{14}$$

Combining Equations (5.63), (5.56) and (5.53), it follows that

$$\left|\frac{n^\circ - n}{n}\right| = \left|\frac{f_5 - f_6}{1 + f_6}\right| \tag{5.64}$$

$$f_5 = \frac{\log(1 + \mu f_3)}{\log\left[\dfrac{\bar{w}(\bar{x}_{*2})}{\bar{w}(\bar{x}_{*1})}\right]} \tag{5.65}$$

$$f_6 = \frac{\log(1 + \mu f_4)}{\log\left[\dfrac{\bar{w}(\bar{x}_{*2})}{\bar{w}(\bar{x}_{*1})}\right]} \tag{5.66}$$

It can be checked that Equations (5.64) to (5.66) lead to

$$\left|\frac{n^\circ - n}{n}\right| \leq c_{15}\mu \tag{5.67}$$

$$c_{15} = \frac{c_{12} + c_{14}}{\log\left[\dfrac{\bar{w}(\bar{x}_{*2})}{\bar{w}(\bar{x}_{*1})}\right]}\left\{1 + \frac{c_{14}}{\log\left[\dfrac{\bar{w}(\bar{x}_{*2})}{\bar{w}(\bar{x}_{*1})}\right]}\right\} > 0 \tag{5.68}$$

Thus, it is enough to have $\mu \leq \varepsilon/c_{15}$ to obtain

$$\left|\frac{n^\circ - n}{n}\right| \leq \varepsilon$$

Proof of step 4: determination of the parameter κ_w. The next parameter to identify is κ_w, using Equation (5.11). An estimate of this parameter can be computed from Equation (5.11) using the following formula:

$$\kappa_w^\circ = \theta^\circ(\bar{x}_{*2}) \sqrt[n^\circ]{\frac{\left[\frac{\mathrm{d}\theta^\circ(\bar{x})}{\mathrm{d}\bar{x}}\right]_{\bar{x}=\bar{x}_*^\circ}}{\left[\frac{\mathrm{d}\theta^\circ(\bar{x})}{\mathrm{d}\bar{x}}\right]_{\bar{x}=\bar{x}_*^\circ} - \left[\frac{\mathrm{d}\theta^\circ(\bar{x})}{\mathrm{d}\bar{x}}\right]_{\bar{x}=\bar{x}_{*2}}}} \tag{5.69}$$

Using Equations (5.54), (5.46) and (5.47), the following equation is obtained for $\bar{w}(\bar{x}_*^\circ) \leq 0$ (a similar relation is obtained for $\bar{w}(\bar{x}_*^\circ) \geq 0$):

$$\begin{aligned}
&\frac{\left[\frac{\mathrm{d}\theta^\circ(\bar{x})}{\mathrm{d}\bar{x}}\right]_{\bar{x}=\bar{x}_*^\circ}}{\left[\frac{\mathrm{d}\theta^\circ(\bar{x})}{\mathrm{d}\bar{x}}\right]_{\bar{x}=\bar{x}_*^\circ} - \left[\frac{\mathrm{d}\theta^\circ(\bar{x})}{\mathrm{d}\bar{x}}\right]_{\bar{x}=\bar{x}_{*2}}} \\
&= \frac{\frac{\kappa_x}{\rho\kappa_w}\vartheta_1(\bar{x}_*^\circ) + 1 + (2\sigma - 1)\,[-\bar{w}(\bar{x}_*^\circ)]^n}{\frac{\kappa_x}{\rho\kappa_w}\,[-\vartheta_1(\bar{x}_{*2}) + \vartheta_1(\bar{x}_*^\circ)] + \bar{w}(\bar{x}_{*2})^n + (2\sigma - 1)\,[-\bar{w}(\bar{x}_*^\circ])^n} \\
&= \frac{1}{\bar{w}(\bar{x}_{*2})^n}\,(1 + \mu f_7)
\end{aligned} \tag{5.70}$$

where

$$|f_7| \leq c_{16} \tag{5.71}$$

$$c_{16} = c_{17} + 2c_{18} + c_{17}c_{18} \tag{5.72}$$

$$c_{17} = \frac{c_{10}\kappa_x}{\rho\kappa_w} + 2\sigma c_8^n \tag{5.73}$$

$$c_{18} = \frac{1}{\bar{w}(\bar{x}_{*2})^n}\left(\frac{2c_{10}\kappa_x}{\rho\kappa_w} + 2\sigma c_8^n\right) \tag{5.74}$$

On the other hand, from Equations (5.39) and (5.62) it follows that

$$\theta^\circ(\bar{x}_{*2}) = \vartheta_2(\bar{x}_{*2}) + \kappa_w\bar{w}(\bar{x}_{*2}) = \kappa_w\bar{w}(\bar{x}_{*2})\,(1 + \mu f_8) \tag{5.75}$$

where

$$|f_8| \leq c_{19}\mu \tag{5.76}$$

$$c_{19} = \frac{c_2}{\kappa_x \bar{w}(\bar{x}_{*2})} \tag{5.77}$$

Combining Equations (5.75), (5.70) and (5.69), the following is obtained through a mathematical analysis:

$$\begin{aligned}\left|\frac{\kappa_w^\circ - \kappa_w}{\kappa_w}\right| &= \left|\bar{w}(\bar{x}_{*2})^{f_9/(1+f_9)} \times (1+\mu f_8)(1+\mu f_7)^{1/[n(1+f_9)]} - 1\right| \\ &\leq c_{20}\mu \end{aligned} \tag{5.78}$$

with

$$|f_9| = \left|\frac{n^\circ - n}{n}\right| \leq c_{15}\mu, \qquad \mu \leq c_{21}$$

and where c_{20} and c_{21} are some positive functions of the parameters

$$\kappa_x,\ \kappa_w,\ \rho,\ \sigma,\ n,\ \bar{q},\ \bar{x}_{*1},\ \bar{x}_{*2}$$

Therefore, it is enough to have $\mu \leq \varepsilon / c_{20}$ to obtain

$$\left|\frac{\kappa_w^\circ - \kappa_w}{\kappa_w}\right| \leq \varepsilon$$

Proof of step 5: determination of the parameter ρ. Identification of the parameter ρ is done using Equation (5.12):

$$\rho^\circ = \frac{\left[\dfrac{\mathrm{d}\theta^\circ(\bar{x})}{\mathrm{d}\bar{x}}\right]_{\bar{x}=\bar{x}_*^\circ}}{X_{\max}\kappa_w^\circ} \tag{5.79}$$

From Equations (5.79) and (5.47), the following equation is obtained for $\bar{w}(\bar{x}_*^\circ) \leq 0$ (a similar relation is obtained for $\bar{w}(\bar{x}_*^\circ) \geq 0$):

$$\begin{aligned}\left|\frac{\rho^\circ - \rho}{\rho}\right| &= \left|\frac{\kappa_x}{\rho\kappa_w(1+f_{10})}\vartheta_1(\bar{x}_*^\circ) - \frac{f_{10}}{1+f_{10}} + \frac{2\sigma - 1}{1+f_{10}}[-\bar{w}(\bar{x}_*^\circ)^n]\right| \\ &\leq c_{22}\mu \end{aligned} \tag{5.80}$$

where

$$f_{10} = \frac{\kappa_w^\circ - \kappa_w}{\kappa_w} \tag{5.81}$$

$$c_{22} = \frac{2\kappa_x c_{10}}{\rho \kappa_w} + 2c_{20} + 4\sigma c_8 \tag{5.82}$$

Thus, it is enough to have $\mu \leq \varepsilon / c_{22}$ to obtain

$$\left| \frac{\rho^\circ - \rho}{\rho} \right| \leq \varepsilon$$

Proof of step 6: determination of the parameter σ. To determine the parameter σ, the function $\bar{w}(\bar{x})$ needs to be computed using Equation (5.13). However, in this equation the parameter κ_w and the function θ are unknown. Thus the computable function is defined as

$$\bar{w}^\circ(\bar{x}) = \frac{\theta^\circ(\bar{x})}{\kappa_w^\circ} \tag{5.83}$$

An estimate of the parameter σ can be computed from Equation (5.14) as follows:

$$\sigma^\circ = \frac{1}{2} \left(\frac{\dfrac{\left[\dfrac{d\bar{w}^\circ(\bar{x})}{d\bar{x}} \right]_{\bar{x}=\bar{x}_{*3}}}{X_{\max} \rho^\circ} - 1}{[-\bar{w}^\circ(\bar{x}_{*3})]^{n^\circ}} + 1 \right) \tag{5.84}$$

where $\bar{x}_{*3}$ is a design parameter such that

$$\bar{w}^\circ(\bar{x}_{*3}) < \frac{\theta^\circ(-1)}{2\kappa_w^\circ} = r_3 < 0$$

It is to be noted that, due to the relations

$$\kappa_w > 0 \qquad \text{and} \qquad \left| \frac{\kappa_w^\circ - \kappa_w}{\kappa_w} \right| \leq \varepsilon$$

it follows that $\kappa_w^\circ > 0$. On the other hand, it has been shown above that the function θ° is strictly increasing, so a unique computable value $-1 < \bar{x}_{*4}$ exists such that $\bar{w}^\circ(\bar{x}_{*4}) = r_3$. Any design parameter

$-1 < \bar{x}_{*3} < \bar{x}_{*4}$ is then appropriate. Using Assumption 2 with Equation (5.83), it can be checked that

$$\bar{w}(\bar{x}_{*3}) < 0 \qquad \text{for } \mu < \frac{\kappa_x |r_3|}{2c_2}$$

Combining Equations (5.83), (5.54) and (5.62) gives

$$\frac{\dfrac{\left[\dfrac{\mathrm{d}\bar{w}^\circ(\bar{x})}{\mathrm{d}\bar{x}}\right]_{\bar{x}=\bar{x}_{*3}}}{X_{\max}\rho^\circ}-1}{[-\bar{w}^\circ(\bar{x}_{*3})]^{n^\circ}} = (2\sigma-1)[-\bar{w}(\bar{x}_{*3})]^{n-n^\circ} \times \left\{ \frac{\dfrac{\kappa_x \vartheta_1(\bar{x}_{*3})}{\rho^\circ\kappa_w^\circ(2\sigma-1)[-\bar{w}(\bar{x}_{*3})]^n} + \dfrac{\rho\kappa_w/(\rho^\circ\kappa_w^\circ)-1}{(2\sigma-1)[-\bar{w}(\bar{x}_{*3})]^n} + \dfrac{\rho\kappa_w}{\rho^\circ\kappa_w^\circ}}{\left[\dfrac{\vartheta_2(\bar{x}_{*3})}{\kappa_w^\circ \bar{w}(\bar{x}_{*3})} + \dfrac{\kappa_w}{\kappa_w^\circ}\right]^{n^\circ}} \right\} \tag{5.85}$$

Denote

$$h = [-\bar{w}(\bar{x}_{*3})]^n \qquad \text{and} \qquad f_{11} = \frac{h^{f_9}-1}{\mu}$$

where it is recalled that

$$f_9 = \frac{n^\circ - n}{n} \qquad \text{and} \qquad |f_9| \leq c_{15}\mu$$

Then, it can be checked that $|f_{11}| \leq c_{26}\mu$, where c_{26} is some positive function of the parameters

$$\kappa_x,\ \kappa_w,\ \rho,\ \sigma,\ n,\ \bar{q},\ \bar{x}_{*1},\ \bar{x}_{*2},\ \bar{x}_{*3}$$

On the other hand,

$$\frac{\dfrac{\kappa_x \vartheta_1(\bar{x}_{*3})}{\rho^\circ\kappa_w^\circ(2\sigma-1)[-\bar{w}(\bar{x}_{*3})]^n} + \dfrac{\rho\kappa_w/(\rho^\circ\kappa_w^\circ)-1}{(2\sigma-1)[-\bar{w}(\bar{x}_{*3})]^n} + \dfrac{\rho\kappa_w}{\rho^\circ\kappa_w^\circ}}{\left[\dfrac{\vartheta_2(\bar{x}_{*3})}{\kappa_w^\circ \bar{w}(\bar{x}_{*3})} + \dfrac{\kappa_w}{\kappa_w^\circ}\right]^{n^\circ}} = 1 + \mu f_{12} \tag{5.86}$$

where

$$|f_{12}| \leq c_{25}$$
$$c_{25} = c_{23} + 2c_{24} + c_{23}c_{24}$$
$$c_{23} = \frac{4\kappa_x c_{10} + 6c_{20} + 4c_{22}}{(2\sigma - 1)\left[-\bar{w}(\bar{x}_{*3})^n\right]} + 6c_{20} + 4c_{22}$$
$$c_{24} = \frac{4c_{13}}{\kappa_w \left|\bar{w}(\bar{x}_{*3})\right|} + 4c_{20}$$

From Equations (5.84) to (5.86), it follows that

$$\left|\frac{\sigma^\circ - \sigma}{\sigma}\right| \leq c_{27}\mu \tag{5.87}$$

$$c_{27} = \frac{2(c_{25} + c_{26}) + c_{25}c_{26}}{4\sigma} \tag{5.88}$$

Thus, it is enough to have $\mu \leq \varepsilon / c_{27}$ to obtain

$$\left|\frac{\sigma^\circ - \sigma}{\sigma}\right| \leq \varepsilon$$

With step 6, the proof of Theorem 5 is ended.

5.2.4 Numerical Simulation Example

In this section the Bouc–Wen model given by the following unknown parameters is considered:

$$\kappa_x = 2, \ \kappa_w = 2, \ \rho = 1, \ \sigma = 3, \ n = 1.5$$

The objective is to use the technique presented in the previous sections to identify its parameters. The 10 steps described in Section 5.2.2 and summarized in Table 5.1 are followed systematically, including the robustness issues treated in Section 5.2.3.

Step 1

The first step in the identification procedure is the choice of the T-periodic input signals. Due to Assumption 2 (Section 5.2.3),

$$\left|\dot{\xi}(\tau)\right| \leq \mu \left|\dot{x}(\tau)\right|$$

This implies that the derivative $\dot{\xi}(\tau)$ of the disturbance $\xi(\tau)$ needs to be zero whenever the derivative of the input signal $x(\tau)$ is zero. Thus, a sine wave input signal candidate would impose the condition that $\dot{\xi}(\tau)$ should be very small around the time instants $0+mT$ and $T/2+mT$ (m is any positive integer), which is unlikely to happen in practice. For this reason, a good choice is a triangular input signal, so that the derivative $\dot{\xi}(\tau)$ needs only to be small with respect to the slope of the input signal, which is constant (in absolute value).

The next design parameter to be chosen is the frequency of the input signal. Since the Bouc–Wen model is rate independent (Section 3.5), its input–output behaviour is independent of the frequency of the input signal. Therefore $T=1$ and $T^{+}=T/2$ are taken and $X_{\max}=-X_{\min}=0.2$ is chosen.

Step 2

In this step, the value $q\neq 0$ must be chosen to obtain a second input signal $x_1(t)=x(t)+q$. The signals $x(t)$ and $x_1(t)$ are given in Figure 5.3 with $q=0.1$.

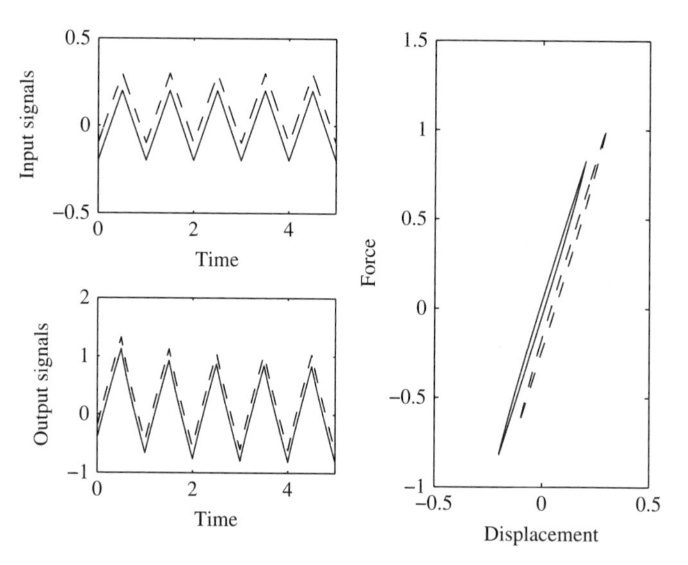

Figure 5.3 Upper left: solid, input signal $x(t)$; dashed, input signal $x_1(t)$. Lower left: solid, output $\Phi_{\mathrm{BW}}(x)(t)$; dashed, output $\Phi_{\mathrm{BW},1}(x)(t)$. Right: limit cycles $(x, \bar{\Phi}_{\mathrm{BW}})$ (solid) and $(x_1, \bar{\Phi}_{\mathrm{BW},1})$ (dashed) that have been obtained for the time interval $[4T, 5T]$.

In practice, the input and output data are in the form of a finite number m of samples $x(kh)$, $\bar{\Phi}(kh)$ where h is the sampling period, $k = 0, 1, \ldots, m$. These samples have to be taken once the output of the system is in a steady state. Note that, since the identification technique uses only the loading part of the limit cycle, the time instant $kh = 0$ can be chosen such that $x(k = 0)$ corresponds to the lowest value of x and the time instant mh so that $x(k = m)$ corresponds to the largest value of x. This implies that the samples that are used for identification purposes verify

$$x(i) < x(i+1) \qquad \text{for all } 0 \leq i < m$$

as the loading part of the limit cycle is being considered.

Step 3

The estimate κ_x° of the coefficient κ_w is computed from Equation (5.5) as

$$\kappa_x^\circ = \frac{\bar{\Phi}_{\mathrm{BW},1}\left(x(0)+q\right) - \bar{\Phi}_{\mathrm{BW}}\left(x(0)\right)}{q} \tag{5.89}$$

where $x(0)$ is the value of x at the time instant $k = 0$. In the absence of noise, then $\kappa_x^\circ = \kappa_w$. In the presence of noise, this is no longer the case; the effect of noise on parameter identification has been studied in the previous section (see Equation (5.35) and subsequent analysis).

Step 4

An estimate $\theta^\circ(x)$ of the function $\theta(x)$ is computed from Equation (5.6) as

$$\theta^\circ\left(x(i)\right) = \bar{\Phi}_{\mathrm{BW}}\left(x(i)\right) - \kappa_x^\circ x(i) \qquad \text{for } i = 0, \ldots, m \tag{5.90}$$

Step 5

It has been shown in the previous section that the estimate $\theta^\circ(x)$ is strictly increasing and has a unique zero, that is a unique point

x_* exists such that $\theta^\circ(x_*) = 0$. Since all the samples $x(i)$ are such that $x(i) < x(i+1)$, then $\theta^\circ(x(i)) < \theta^\circ(x(i+1))$. The existence and unicity of the zero of the function θ° shows that a unique integer r exists such that $\theta^\circ(x(r)) \leq 0 < \theta^\circ(x(r+1))$. This implies that $x(r) \leq x_* < x(r+1)$, and a linear interpolation gives an estimate x_*° of the zero x_*. A simple computer program can be done to determine the integer r.

Step 6

An estimate of the parameter a is computed from Equation (5.8) as follows:

$$a^\circ = \frac{\theta^\circ(x(r+1)) - \theta^\circ(x(r))}{x(r+1) - x(r)} \tag{5.91}$$

Step 7

Choosing the design parameters

$$x_{*2} = x(l_2) > x_{*1} = x(l_1) > x_*^\circ$$

the estimates n° and b° are computed from Equations (5.9) and (5.10) as follows:

$$n^\circ = \frac{\log\left[\dfrac{\dfrac{\theta^\circ(x(l_2+1)) - \theta^\circ(x(l_2))}{x(l_2+1) - x(l_2)} - a^\circ}{\dfrac{\theta^\circ(x(l_1+1)) - \theta^\circ(x(l_1))}{x(l_1+1) - x(l_1)} - a^\circ}\right]}{\log\left[\dfrac{\theta^\circ(x_{*2})}{\theta^\circ(x_{*1})}\right]} \tag{5.92}$$

$$b^\circ = \frac{a^\circ - \dfrac{\theta^\circ(x(l_2+1)) - \theta^\circ(x(l_2))}{x(l_2+1) - x(l_2)}}{\theta^\circ(x_{*2})^{n^\circ}} \tag{5.93}$$

Step 8

Estimates of the parameters κ_w and ρ are computed from Equations (5.11) and (5.12) as follows:

$$\kappa_w^\circ = \sqrt[n^\circ]{\frac{a^\circ}{b^\circ}} \tag{5.94}$$

$$\rho^\circ = \frac{a^\circ}{\kappa_w^\circ} \tag{5.95}$$

Step 9

An estimate of the function $\bar{w}(x)$ is computed from Equation (5.13) in the form

$$\bar{w}^\circ\,(x(i)) = \frac{\theta^\circ\,(x(i))}{\kappa_w^\circ} \qquad \text{for } i = 0, \ldots, m \tag{5.96}$$

Step 10

Choose a design parameter $x_{*3} = x(l_3) < x_*^\circ$. Then an estimate of the parameter σ is computed from Equation (5.14) as

$$\sigma^\circ = \frac{1}{2}\left(\frac{\dfrac{\dfrac{\bar{w}^\circ\,(x(l_3+1)) - \bar{w}^\circ\,(x(l_3))}{x(l_3+1) - x(l_3)} - 1}{\rho^\circ}}{[-\bar{w}^\circ(x_{*3})]^{n^\circ}} + 1 \right) \tag{5.97}$$

The numerical simulation gives the final result:

$$\kappa_x^\circ = 2.0000,\ \kappa_w^\circ = 2.0059,\ \rho^\circ = 0.9971,\ n^\circ = 1.4954,\ \sigma^\circ = 2.9728$$

5.3 MODELLING AND IDENTIFICATION OF A MAGNETORHEOLOGICAL DAMPER

In this section the case study of a magnetorheological (MR) damper is considered. The results of the previous chapters are used to gain insight into some existing models for this nonlinear device. In particular, the identification technique of the previous sections is appropriately modified to determine the parameters of the MR damper.

5.3.1 Some Insights into the Viscous + Bouc–Wen Model for Shear Mode MR Dampers

In this section the model is analysed of a shear mode MR damper proposed in Reference [90]. A physical description of this damper will be done in Section 5.4.1. For the moment, it is viewed as a nonlinear system whose inputs are the displacement of the damper $x(t)$ and the voltage $v(t)$ at the level of the coil; and whose output is the force $F(t)$ applied by the damper (see Figure 5.4).

The output force is related to its input displacement and voltage as follows:

$$F(x)(t) = k(v)\dot{x}(t) + \alpha(v)z(t) \tag{5.98}$$

$$\dot{z} = A\dot{x} - \beta|\dot{x}|\,|z|^{n-1}z - \gamma\dot{x}|z|^{n} \tag{5.99}$$

where the parameters k and α are voltage dependent. This model is represented in Figure 5.5 as the sum of a viscous friction term $k(v)\dot{x}(t)$ and a hysteresis contribution $\alpha(v)z(t)$, where the state $z(t)$ obeys the standard Bouc–Wen differential equation (5.99). Note that the only difference between the model (5.98)–(5.99) and the model (2.4)–(2.5) is the term $\dot{x}$ instead of x in Equation (5.98). It is clear that the only corresponding changes in Theorem 3 are obtained by putting $\dot{x}$ instead of x in Equations (3.40) and (3.41).

Let us first consider that the input voltage is zero (the conclusions of this section are similar to those for the nonzero voltage input). In

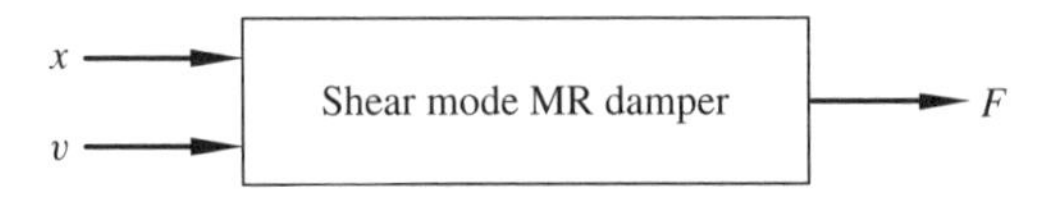

Figure 5.4 Input–output representation of the MR damper.

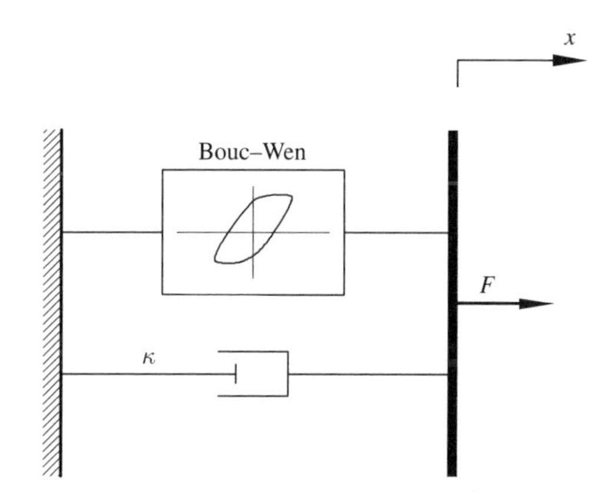

Figure 5.5 Mechanical model of the MR damper.

this case, the following values have been taken for the Bouc–Wen model parameters:

$$A = 120, \quad \beta = 300\,\text{cm}^{-1}, \quad \gamma = 300\,\text{cm}^{-1}, \quad k = 0.032\,\text{N}\cdot\text{s/cm},$$
$$\alpha = 27.3\,\text{N/cm}, \quad n = 1 \tag{5.100}$$

To check the validity of the model, the same input signal has been applied to the experimental MR damper and to its model with the initial condition $z(0) = 0$. A reasonable matching has been observed between the force applied by the real MR damper (experimental data) and the force calculated using Equations (5.98) and (5.99) (numerically obtained) [90]. In the following some insights into this model are presented.

It has been observed in Chapter 3 that the standard Bouc–Wen model is overparametrized, that is the model contains more parameters than necessary. In particular, it can be checked that for any positive scalar a, the Bouc–Wen model given by the set of parameters

$$k' = k, \quad \alpha' = a\alpha, \quad A' = A/a, \quad \beta' = \beta, \quad \gamma' = \gamma, \quad n' = n \tag{5.101}$$

will have an input–output behaviour equal to that of the model (5.98)–(5.99) with the set of parameters (5.100). To illustrate this point, a numerical simulation is used where $a = 10$. The same input signal is applied to the model (5.98)–(5.99) with the two sets of parameters (5.100) and (5.101) and with the initial condition $z(0) = 0$. Figure 5.6 shows that the obtained forces are exactly equal.

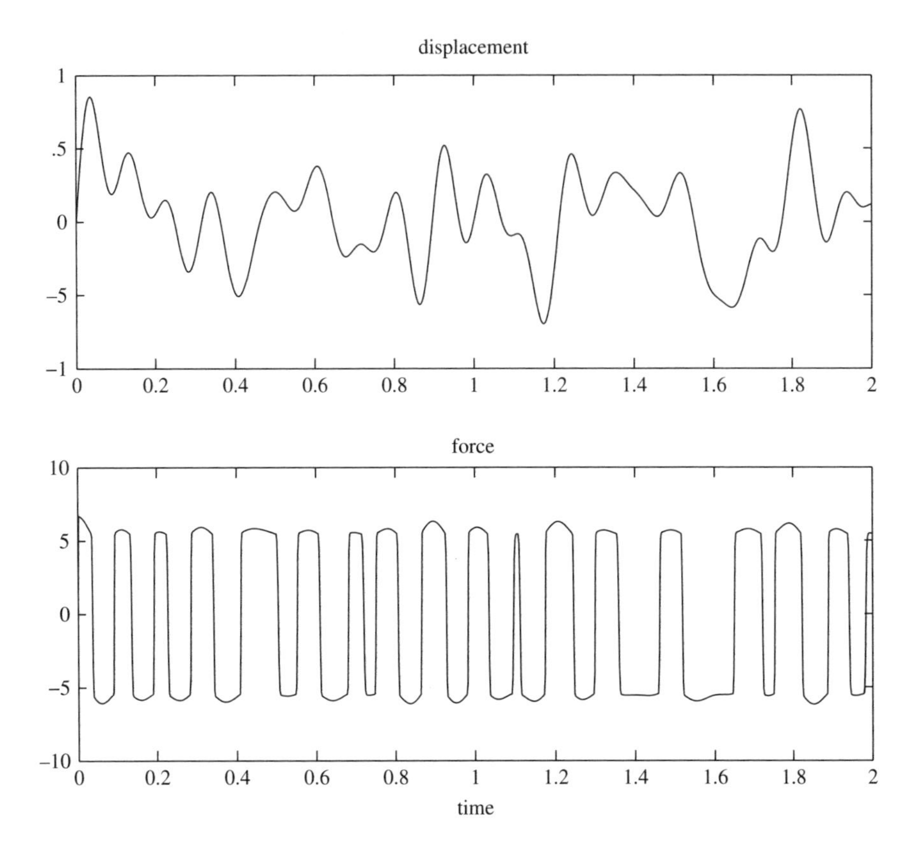

Figure 5.6 Upper: input signal. Lower: response of the Bouc–Wen model (5.98)–(5.99) to the two sets of parameters (5.100) and (5.101).

Note that the parameters β and γ in (5.100) have been chosen so that $\beta = \gamma$. This choice, common in the Bouc–Wen model literature (see Reference [62] for example), is often used either to simplify the determination of the set of parameters $\{k, \alpha, A, \beta, \gamma, n\}$ from experimental data or because of physical considerations [133, page 14]. However, this choice does not eliminate the overparametrization of the Bouc–Wen model. Indeed, in this case, the set of parameters that describe the standard model (5.98)–(5.99) is $\{k, \alpha, A, \beta = \gamma, n\}$, which has five parameters. On the other hand, Equations (3.11) show that, with this choice, $\sigma = \beta/(\beta + \gamma) = 1/2$ so that the corresponding set of normalized parameters is $\{\kappa_x, \kappa_w, \rho, n\}$, which has only four parameters. This means that, if the standard Bouc–Wen model with the condition $\beta = \gamma$ is used, the rest of the parameters can be determined from experimental data, and five parameters will need to be determined. If instead the normalized Bouc–Wen model with the equivalent condition $\sigma = 1/2$ is used, only four parameters will need to be determined. In other words, trying to impose some relations between the standard Bouc–Wen model parameters does not necessarily eliminate the overparametrization. To eliminate it, a sound way is to use the normalized version of the Bouc–Wen model:

$$F(x)(t) = \kappa_x(v)\dot{x}(t) + \kappa_w(v)w(t) \tag{5.102}$$

$$\dot{w}(t) = \rho\left(\dot{x}(t) - \sigma|\dot{x}(t)||w(t)|^{n-1}w(t) + (\sigma - 1)\dot{x}(t)|w(t)|^n\right) \tag{5.103}$$

where

$$\kappa_x(v) = k(v) \quad \text{and} \qquad \kappa_w(v) = z_0\alpha(v)$$

The other normalized Bouc–Wen model parameters are given by the relations (3.11).

Now, consider the hysteresis loop obtained using the model (5.98)–(5.99) with the set of parameters (5.100) and with the input

$$x(t) = \sin(t)$$

The part of the hysteresis loop corresponding to $\alpha(v)z(t)$ is displayed in Figure 5.7. It shows that, after a transient, the obtained steady state loop presents a sharp transition at the velocity sign change. Then, on both loading and unloading, the value of the hysteresis term of the output force is practically constant. This behaviour is difficult to predict from the values of the parameters (5.100) due to the lack in the literature of an analytical study of the relationship between the Bouc–Wen model parameters and the shape of the hysteresis loop. However, Chapter 4 has used the normalized version of the Bouc–Wen model to analyse the effect of the normalized parameters on the limit cycle. Thus, to explain the shape of the hysteresis loop in Figure 5.7, the normalized model that corresponds to the standard Bouc–Wen model (5.98)–(5.99) needs to be determined.

Using the relations (3.11), the set of the corresponding normalized parameters is

$$\kappa_x = 0.032 \text{ N s/cm}, \ \kappa_w = 5.46\,\text{N}, \ \rho = 600\,\text{cm}^{-1}, \ \sigma = 0.5, \ n = 1 \tag{5.104}$$

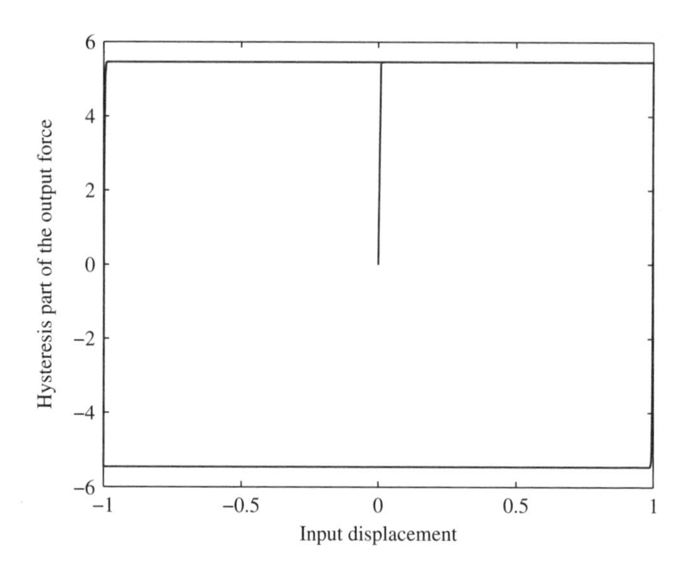

Figure 5.7 Hysteresis loop corresponding to the part $\alpha(v)z(t)$ of the Bouc–Wen model (5.98)–(5.99) with the set of parameters (5.100) to the input displacement $x(t) = \sin(t)$.

In Reference [90], the input displacement varies typically between $-1\,\text{cm}$ and $1\,\text{cm}$ so that its maximal value is $X_{\max} = 1\,\text{cm}$. On the other hand, Section 4.6 shows that the main slope of the linear region of the limit cycle is $\rho X_{\max} = 600$. Note that this slope has a large value (with respect to unity), so that the linear region in Figure 4.3 will be almost vertical, as observed in Figure 5.7.

It will now be demonstrated that the abscissa of the point P_{tp} of Figure 4.3, where the plastic region starts, is almost equal to -1. Combining Equations (4.92) and (4.18) along with the fact that $\rho X_{\max}$ is large with respect to unity, the abscissa of the point P_{tp} is obtained as

$$\bar{x}_{tp} \approx -1 + \frac{1}{\rho X_{\max}} \left[\varphi^{+}_{\sigma,n}\left(\sqrt[n]{1 - r_2}\right) - \varphi^{+}_{\sigma,n}(-1) \right] \tag{5.105}$$

In Equation (5.105) the terms between brackets are independent of $\rho X_{\max}$, so that

$$\lim_{\rho X_{\max} \to \infty} \bar{x}_{tp} = -1$$

This means that, for large values of $\rho X_{\max}$, the abscissa $\bar{x}_{tp}$ of the point P_{tp} where the plastic region starts is almost -1. This fact, along with the symmetry of the Bouc–Wen model hysteresis loop, explains the sharp transition at the velocity sign change in Figure 5.7 and the fact that the hysteresis term of the force remains constant during the loading and unloading.

Note that, in Equation (5.105), the fact that $\bar{x}_{tp} \to (-1)$ as $\rho \to \infty$ is independent of the particular values of the parameters n and σ. This suggests that, if $\rho X_{\max}$ is large with respect to unity, then for a wide range of the parameters n and σ, the hysteresis loop of the corresponding Bouc–Wen model will be similar. To illustrate this point, consider the following set of normalized parameters:

$$\kappa_x = 0.032 \text{ N s/cm}, \;\; \kappa_w = 5.46\,\text{N}, \;\; \rho = 600\,\text{cm}^{-1}, \;\; \sigma = 10, \;\; n = 10 \tag{5.106}$$

This set is the same as (5.104) except for the values of the parameters n and σ. The corresponding normalized Bouc–Wen models are excited with a random input signal whose frequency content covers the interval [0,10 Hz], which is a range of frequencies that is common for some civil engineering structures. Figure 5.8 gives the obtained hysteresis terms of the output forces in both cases. It can be seen that both responses are practically equal.

The conclusion of this analysis is that, for large values of $\rho X_{\max}$, the values of the parameters n and σ are practically irrelevant in the sense that any value of these parameters will lead to almost the same behaviour of the corresponding Bouc–Wen model.

5.3.2 Alternatives to the Viscous + Bouc–Wen Model for Shear Mode MR Dampers

The previous section has highlighted two characteristics of the shear mode MR damper model proposed in Reference [90]:

1. The overparametrization of the associated Bouc–Wen model.
2. The large value of the quantity $\rho X_{\max}$.

The overparametrization of the Bouc–Wen model is due to the use of this model under its standard form (5.98)–(5.99). This form uses a set of six parameters $\{k, \alpha, A, \beta, \gamma, n\}$ to describe the relation between the input displacement and the output restoring force. It has

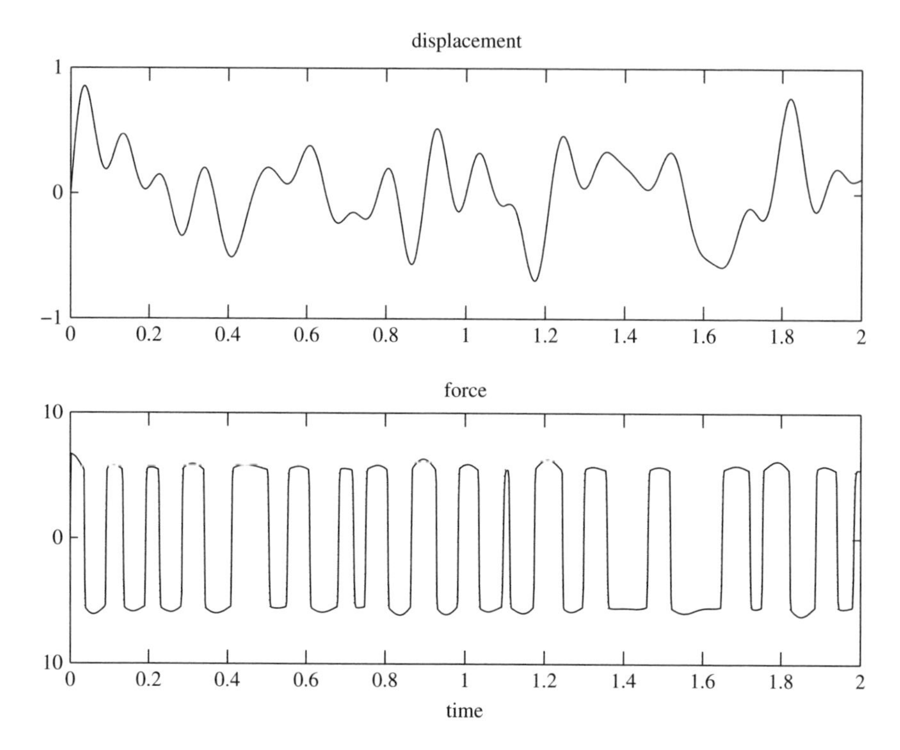

Figure 5.8 Response of the normalized Bouc–Wen model (5.102)–(5.103) to a random input signal with a frequency content within the interval [0,10 Hz]: solid, set of parameters (5.104); dotted, set of parameters (5.106).

been shown in the previous sections that the standard form of the Bouc–Wen model is equivalent to its normalized form, which uses a set of only five parameters $\{\kappa_x, \kappa_w, \rho, \sigma, n\}$. This means that it is more appropriate to use the Bouc–Wen model under its normalized form (5.102)–(5.103).

On the other hand, it has been shown in Reference [90] that the considered shear mode MR damper is described with a reasonable precision using the standard Bouc–Wen based model (5.98)–(5.99) along with the values of the parameters (5.100). The corresponding normalized model (5.102)–(5.103) has been determined with the corresponding values of the parameters (5.104). It has been noted that the term $\rho X_{\max}$ has a large value with respect to unity, which implies that the particular values of the parameters n and σ are irrelevant. This means that the information about the behaviour of the MR damper is captured in at most three parameters κ_x, κ_w and ρ. In other terms, the initial set $\{k, \alpha, A, \beta, \gamma, n\}$

of the standard Bouc–Wen model parameters contains at least three parameters that are irrelevant for the description of the damper.

For the reasons explained in the previous paragraphs, there is a clear need to develop a simplified model for the damper that has a minimal number of parameters which are all relevant. Figure 5.7 suggests the following model:

$$F(t) = \kappa_x(v)\dot{x}(t) + \kappa_w(v)F_C(\dot{x}) \tag{5.107}$$

where κ_x and κ_w are constants that may be voltage dependent and F_C is the Coulomb model for dry friction defined as [134]

$$F_C(\dot{x}) = 1 \qquad \text{for } \dot{x} > 0 \tag{5.108}$$

$$F_C(\dot{x}) = -1 \qquad \text{for } \dot{x} < 0 \tag{5.109}$$

At zero velocity, the Coulomb friction can take any value in the interval $[-1, 1]$. The graph of the Coulomb friction is plotted in Figure 5.9, and it can be seen that it is a static model when the input is taken to be the velocity. The graph of the proposed simplified model (5.107)–(5.109) for the shear MR damper is given in Figure 5.10. It consists of the sum of a viscous friction contribution and a dry friction term. This is also a static model when the considered input is the velocity $\dot{x}$. Note that this model has only two parameters κ_x

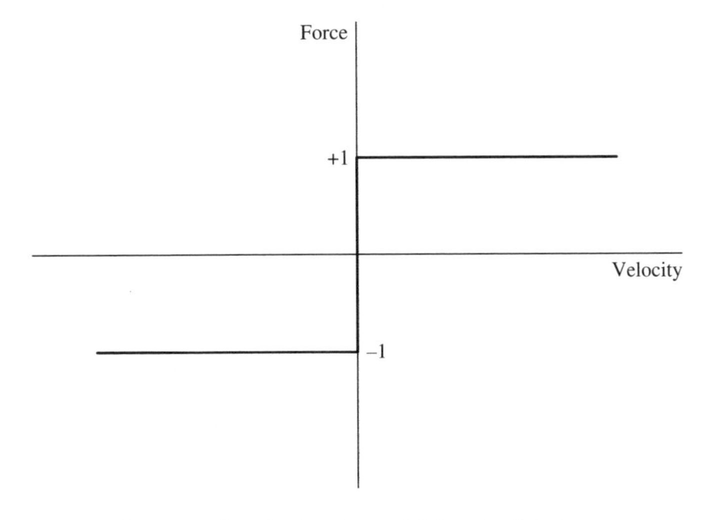

Figure 5.9 Coulomb model for dry friction.

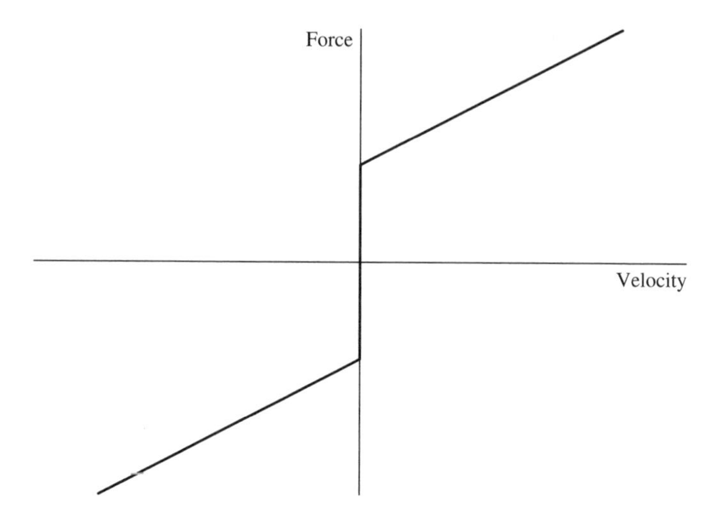

Figure 5.10 Viscous + Coulomb model for the shear mode MR damper (also called the Bingham model [136]).

and κ_w, which is particularly useful for the analysis of the model behaviour and for the design of control laws [135].

In the following, the behaviour is compared of the shear mode MR damper when it is described by the standard Bouc–Wen model (5.98)–(5.99) with the set of parameters (5.100) (or equivalently the normalized Bouc–Wen model (5.102)–(5.103) with the set of parameters (5.104)), and when it is described by the viscous + Coulomb friction model (5.107) with the set of parameters

$$\kappa_x = 0.032\,\text{N s/cm}, \quad \kappa_w = 5.46\,\text{N} \tag{5.110}$$

Figure 5.11 gives the responses of both models to the same input signal. It can be seen that there is a reasonable matching between the obtained forces, which means that the MR damper may be described by a viscous + Coulomb friction model. This fact leads to a reinterpretation of the behaviour of the MR damper.

An MR fluid consists of iron particles suspended in a carrier liquid and responds to a magnetic field with an important change in rheological behaviour. At a microscopical level, the reorientation of the iron particles under the action of the magnetic field induces a change in the stiffness of the material. Since the phenomenon of hysteresis often accompanies the magnetic-dependent characteristics

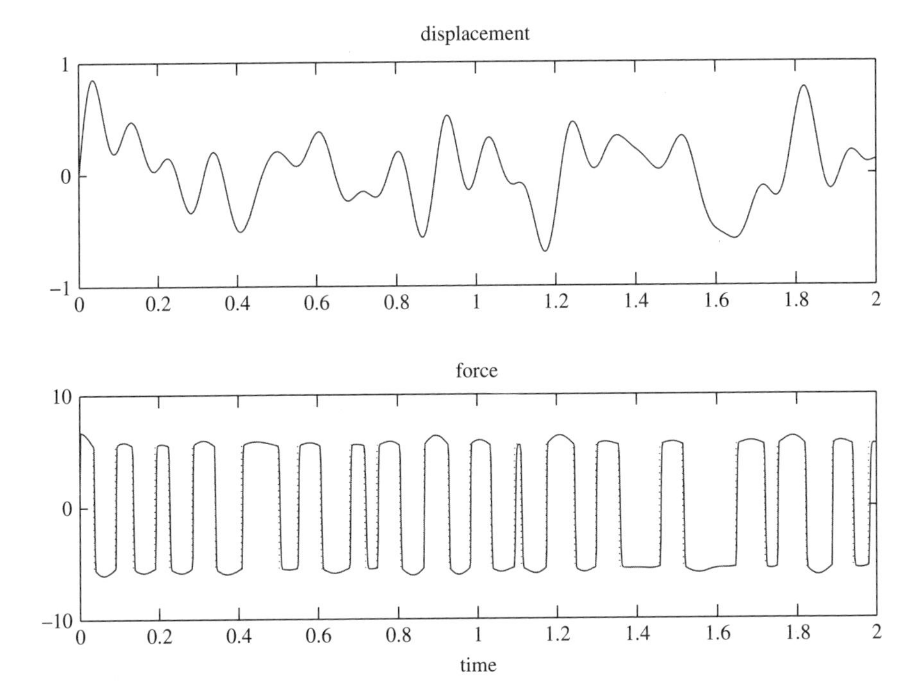

Figure 5.11 Response to a random input signal with a frequency content that covers the interval [0,10 Hz]: solid, standard Bouc–Wen model (5.98)–(5.99) with the set of parameters (5.100); dotted, viscous + Coulomb model (5.107) with the set of parameters (5.110).

in materials, the hysteresis Bouc–Wen model has often been used to describe the behaviour of MR dampers [42]. However, what interests the control engineer and the civil engineer is the macroscopic behaviour of the damper. The viscous + dry friction model of the shear mode MR damper shows that it behaves as a frictional device that has voltage-dependent characteristics of both the viscous and dry friction.

Now, although the viscous + dry friction model may be useful as a tool for the analysis and control of systems that use shear mode MR dampers, it is not appropriate for numerical simulations. Indeed, due to the discontinuity of the Coulomb model for dry friction at the zero velocity, a numerical instability results when the velocity of the damper is close to zero. An alternative to the Coulomb model for dry friction is the Dahl model [138], which is widely used in the literature devoted to friction, both for simulation and control [135,139]. This model consists of a first-order nonlinear differential equation that relates the velocity $\dot{x}$ of the device to the dry frictional force F_{d} in the following way:

$$F_{\mathrm{d}}(t) = \sigma_0 z_{\mathrm{d}}(t) \tag{5.111}$$

$$\dot{z}_{\mathrm{d}} = \dot{x} - \frac{\sigma_0}{F_{\mathrm{c}}} z_{\mathrm{d}} \left|\dot{x}\right| \tag{5.112}$$

where σ_0 and F_{c} are constants, and z_{d} is a state variable.

It is to be noted that the Dahl model for dry friction can be written as a Bouc–Wen model with some appropriate parameters. Indeed, taking

$$\sigma = 1, \quad n = 1, \quad \rho = \frac{\sigma_0}{F_{\mathrm{c}}}, \quad \kappa_w = F_{\mathrm{c}}, \quad w(t) = \frac{\sigma_0}{F_{\mathrm{c}}} z_{\mathrm{d}}(t) \tag{5.113}$$

the Dahl model can be written as

$$F_{\mathrm{d}}(t) = \kappa_w w(t) \tag{5.114}$$

$$\dot{w} = \rho \left(\dot{x} - \left|\dot{x}\right| w\right) \tag{5.115}$$

Thus, the viscous + dry friction model for the shear mode MR damper can be taken in the form

$$F(t) = \kappa_x(v)\dot{x}(t) + \kappa_w(v)w(t) \tag{5.116}$$

$$\dot{w} = \rho \left(\dot{x} - \left|\dot{x}\right| w\right) \tag{5.117}$$

where the constants κ_x and κ_w may be voltage dependent (see Figure 5.12).

The behaviour of the MR damper standard Bouc–Wen model (5.98)–(5.99) with the set of parameters (5.100) is now compared with that of the Dahl model (5.116)–(5.117) with the set of parameters

$$\kappa_x = 0.032\,\mathrm{N\ sec/cm}, \quad \kappa_w = 5.46\,\mathrm{N}, \quad \rho = 600\,\mathrm{cm}^{-1} \tag{5.118}$$

The set (5.118) is the same as (5.104) except for the value of the parameter $\sigma = 1$.

Figure 5.13 gives the responses of both models to the same input signal. It can be seen that the responses are practically equal. This means that the viscous + Dahl model represents well the behaviour

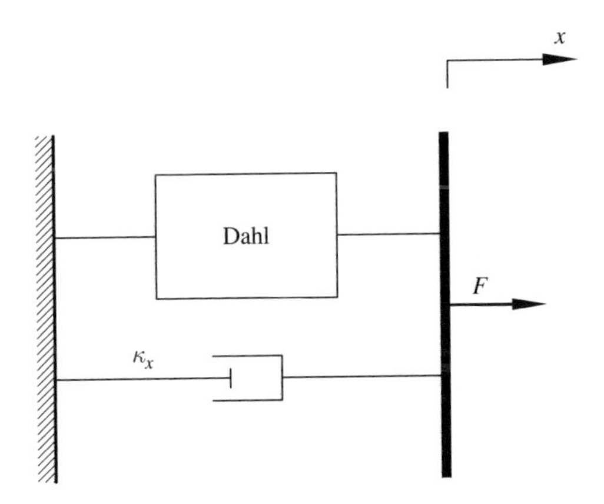

Figure 5.12 Viscous + Dahl model for the MR damper.

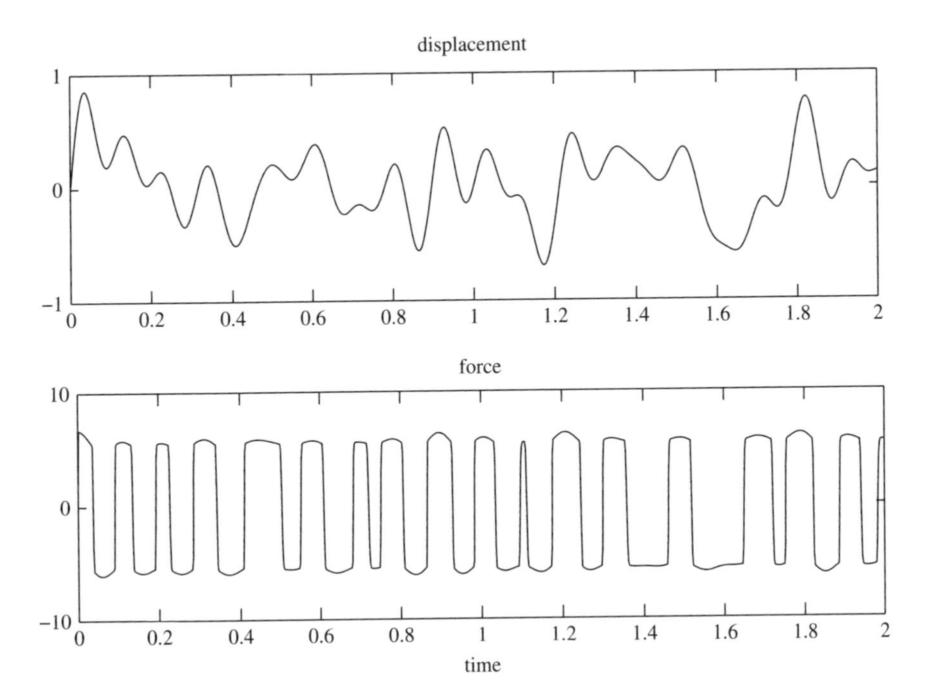

Figure 5.13 Response to a random input signal with a frequency content that covers the interval [0,10 Hz]: solid, normalized Bouc–Wen model (5.102)–(5.103) with the set of parameters (5.104) (or equivalently the standard Bouc–Wen model (5.98)–(5.99) with the set of parameters (5.100)); dotted, viscous + Dahl model (5.116)–(5.117) with the set of parameters (5.118).

of the shear mode MR damper. The next section describes the application of the identification methodology developed in this chapter to determine the parameters of this model.

5.3.3 Identification Methodology for the Viscous + Dahl Model

The Dahl model is a particular case of the Bouc–Wen model. This means that all the results obtained for the Bouc–Wen model are valid for the Dahl model. In particular, when the displacement input signal has a loading–unloading shape, the description of the steady state equations of the output force is given by Theorem 3 in Section 3.5. The identification method assumes the knowledge of the limit cycle, that is the graph $(x(\tau), \bar{F}(\tau))$ parameterized by the variable $\tau \in [0, T]$. Thanks to the symmetry property of this graph, only its loading part will be considered for identification purposes, so that $\tau \in [0, T^+]$ will be taken. In this case, the equation of the loading part of the limit cycle is obtained from Theorem 3, by eliminating the parameter τ in Equation (3.42), in the form

$$\bar{w}(x) = \psi^+_{\sigma,n}\left(\varphi^+_{\sigma,n}\left[-\psi_{\sigma,n}\left(\rho\left(X_{\max} - X_{\min}\right)\right)\right] + \rho\left(x - X_{\min}\right)\right) \quad (5.119)$$

Then, considering that $\sigma = 1$ and $n = 1$, the following derivative is found using Equations (4.53) and (4.54):

$$\frac{d\bar{w}(x)}{dx} = \rho\left(1 - \bar{w}(x)\right) \quad (5.120)$$

Since, by Theorem 3, $-1 < \bar{w}(x) < 1$, it follows from Equation (5.120) that

$$\frac{d\bar{w}(x)}{dx} > 0$$

so that the function $\bar{w}(x)$ is increasing. Again, Theorem 3 shows that the minimal value of $\bar{w}(x)$ is obtained for $x = X_{\min}$ (or equivalently for $\tau = 0$) and the maximal value of $\bar{w}(x)$ is obtained for $x = X_{\max}$ (or equivalently for $\tau = T^+$). These minimal and maximal values are, respectively,

$$-\psi_{1,1}\left(\rho\left(X_{\max} - X_{\min}\right)\right) \qquad \text{and} \qquad \psi_{1,1}\left(\rho\left(X_{\max} - X_{\min}\right)\right)$$

On the other hand, the steady state output force is given by Theorem 3 as

$$\bar{F}(\tau) = \kappa_x \dot{x}(\tau) + \kappa_w \bar{w}(\tau), \qquad \text{where } \tau \in [0, T^+] \tag{5.121}$$

Now, consider a wave T-periodic signal $x(\tau)$. Then, the values of the force at instants $\tau = 0$ and $\tau = T^+$ are, respectively,

$$\bar{F}(0) = \kappa_x \dot{x}(0) - \psi_{1,1}\left(\rho\left(X_{\max} - X_{\min}\right)\right)$$
$$\bar{F}(T^+) = \kappa_x \dot{x}(T^+) + \psi_{1,1}\left(\rho\left(X_{\max} - X_{\min}\right)\right)$$

Thus, the constant κ_x can be determined as

$$\kappa_x = \frac{\bar{F}(0) + \bar{F}(T^+)}{\dot{x}(0) + \dot{x}(T^+)} \tag{5.122}$$

In practice, the input signal has to be chosen in such a way, as to avoid a division by zero in Equation (5.122).

Since κ_x has been determined, the quantity $\kappa_w \bar{w}(\tau)$ can be computed from Equation (5.116) in the form

$$\kappa_w \bar{w}(\tau) = \bar{F}(\tau) - \kappa_x \dot{x}(\tau) \triangleq \theta(\tau) \tag{5.123}$$

Now the function $\theta(\tau)$ with $\tau \in [0, T^+]$ is known. Seen as a function of the variable x, Equation (5.123) shows that θ verifies

$$\kappa_w \bar{w}(x) = \theta(x) \tag{5.124}$$

Then, Equation (5.120) can be written as

$$\frac{\mathrm{d}\theta(x)}{\mathrm{d}x} = a - \rho\,\theta(x) \tag{5.125}$$

where $a = \rho\kappa_w$. The parameter a is determined as

$$a = \left[\frac{\mathrm{d}\theta(x)}{\mathrm{d}x}\right]_{x=x_*} \tag{5.126}$$

where x_* satisfies the relation $\theta(x_*) = 0$. The existence and uniqueness of the zero of the function θ comes from Equation (5.124) and the fact

Table 5.2 Procedure for the identification of the viscous + Dahl model

Step 1	Excite the MR damper with a wave periodic signal $x(t)$. After a transient, the output $F(t)$ will have a steady state $\bar{F}(\tau)$ as proved in Theorem 3.
Step 2	Compute the coefficient κ_x using Equation (5.122).
Step 3	Compute the function θ using Equation (5.123) and determine its zero x_*.
Step 4	Compute the coefficient a using Equation (5.126).
Step 5	Choose a value $x_{*1} > x_*$ and compute the parameters ρ and κ_w using Equations (5.127)) and (5.128)), respectively.

that $\bar{w}(x)$ is increasing from the negative value $-\psi_{1,1}\ (\rho\,(X_{max} - X_{min}))$ to the positive value $\psi_{1,1}\ (\rho\,(X_{max} - X_{min}))$. Now, take some value $x_{*1} > x_*$ and then the parameter ρ can be determined from Equation (5.125) in the form

$$\rho = \frac{a - \left[\dfrac{d\theta(x)}{dx}\right]_{x=x_{*1}}}{\theta\,(x_{*1})} \tag{5.127}$$

Then the parameter κ_w is determined as follows:

$$\kappa_w = \frac{a}{\rho} \tag{5.128}$$

The application of this identification methodology is summarized in Table 5.2.

5.3.4 Numerical Simulations

The considered MR damper is a prototype device shown schematically in Figure 5.14. This experimental device was obtained from the Lord Corporation for testing and evaluation. The device consists of two steel parallel plates. The dimensions of the device are 4.45 cm × 1.9 cm × 2.5 cm. The magnetic field produced in the device is generated by an electromagnet consisting of a coil at one end of the device. Forces are generated when the moving plate, coated with a thin foam saturated with MR fluid, slides between the two parallel plates. The outer plates of the MR device are 0.635 cm apart, and the force capacity of the device is dependent on the strength of the fluid and

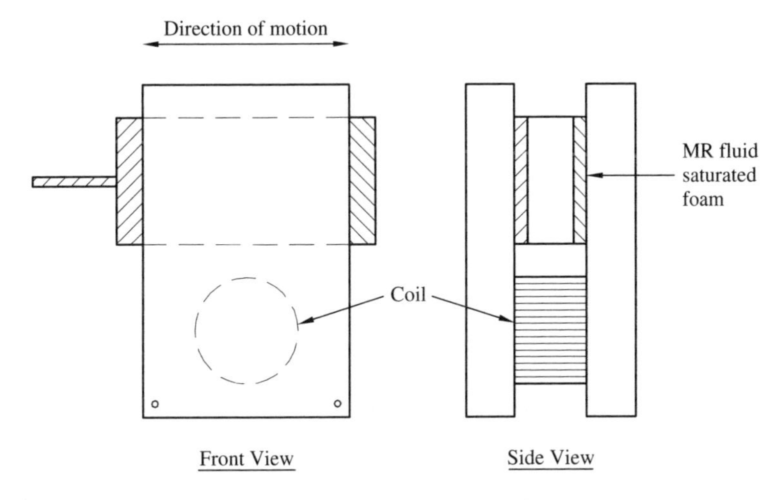

Figure 5.14 Shear mode MR damper.

on the size of the gap between the side plates and the centre plate. Power is supplied to the device by a regulated voltage power supply driving a DC to pulse-width modulator (PWM).

The experimental investigations presented in Reference [90] were performed in the Washington University Structural Control and Earthquake Engineering Laboratory. They have shown that this MR damper can be described with reasonable precision by the standard Bouc–Wen model (5.98)–(5.99) with the following set of parameters:

$$\begin{aligned} &A = 120, \quad \beta = 300\,\mathrm{cm}^{-1}, \quad \gamma = 300\,\mathrm{cm}^{-1}, \quad k_a = 0.032\,\mathrm{N\ s/cm}, \\ &k_b = 0.02\,\mathrm{N\ sec/(cm\ V)}, \quad \alpha_a = 27.3\,\mathrm{N/cm}, \\ &\alpha_b = 26.5\,\mathrm{N/(cm\ V)}, \quad n = 1 \end{aligned} \tag{5.129}$$

and

$$\begin{aligned} \alpha(v) &= \alpha_a + \alpha_b v \\ k(v) &= k_a + k_b v \\ \dot{v} &= -\eta\,(v - u) \end{aligned} \tag{5.130}$$

In this model, the variable u is the command voltage applied to the PWM circuit and η is a positive constant.

In this section, the system under consideration is described by Equations (5.98) and (5.99) along with the relations (5.129) and (5.130). The measured variables are:

- the voltage v,
- the displacement $x(t)$ of the damper and
- the force $F(t)$ delivered by the MR device.

The objective of this section is to use the viscous + Dahl model to describe the MR damper and to use the methodology of Section 5.3.3 to determine its parameters.

The new model of the shear mode MR damper is given by Equations (5.116) and (5.117), where

$$\kappa_x(v) = \kappa_{xa} + \kappa_{xb} v \tag{5.131}$$

$$\kappa_w(v) = \kappa_{wa} + \kappa_{wb} v \tag{5.132}$$

$$\dot{v} = -\eta\,(v - u) \tag{5.133}$$

The unknown parameters to be identified are the following:

$$\kappa_{xa},\ \kappa_{xb},\ \kappa_{wa},\ \kappa_{wb},\ \rho \text{ and } \eta$$

Since the system (5.133) is linear, the parameter η can be determined using standard methods if u is available for measurement. For this reason, in the following the focus is exclusively on the determination of the rest of parameters.

Let u be a constant command voltage. Then, in the steady state, it is clear from Equation (5.133) that $u = v$. It is considered that $v = 0$ so that, from Equations (5.131) and (5.132),

$$\kappa_x(v) = \kappa_{xa} \qquad \text{and } \kappa_w(v) = \kappa_{wa}$$

The same identification technique can be applied for the the case of nonzero constant command voltage. The identification methodology of Section 5.3.3 is applied following the five steps of Table 5.2. The results are given below.

Step 1

The MR damper model (5.98)–(5.99), along with the relations (5.129) and (5.130), is excited by a displacement input signal that is

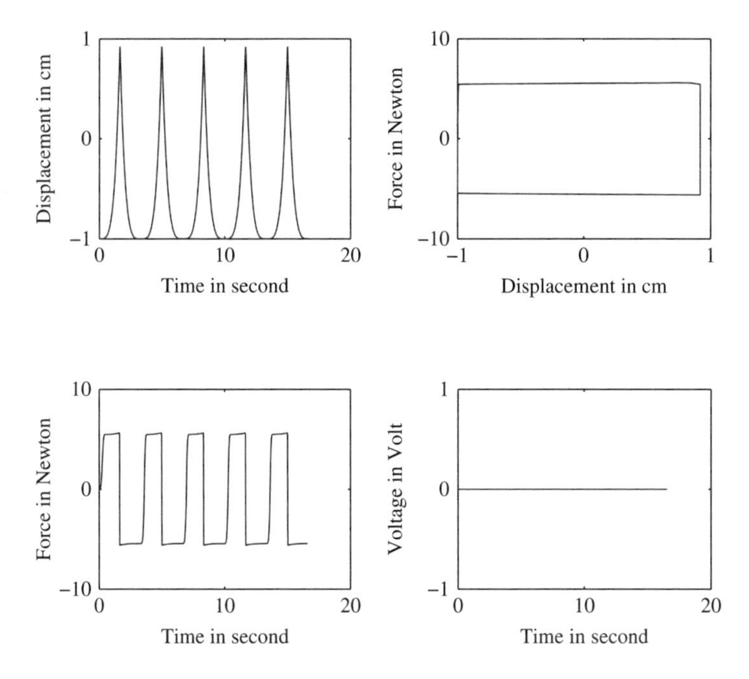

Figure 5.15 Response of the MR damper model.

wave periodic. This signal and the corresponding output force are given in Figure 5.15. Note that the maximal value of the input displacement signal is $X_{\max} = 1\,\text{cm}$.

Step 2

To compute the parameter κ_{xa}, Equation (5.122) is used. The values of the different terms arising in this equation are

$$\begin{aligned} &\bar{F}(0) = -5.4600\,\text{N}, \quad \bar{F}(T^{+}) = 5.6124\,\text{N}, \\ &\dot{x}(0) = 1.1000 \times 10^{-6}\,\text{cm/s}, \quad \dot{x}(T^{+}) = 4.7614\,\text{cm/s} \end{aligned} \tag{5.134}$$

so that the following value is obtained:

$$\kappa_{xa} = 0.0320\,\text{N s/cm}$$

Note that this is equal to the theoretical value obtained in Equation (5.104).

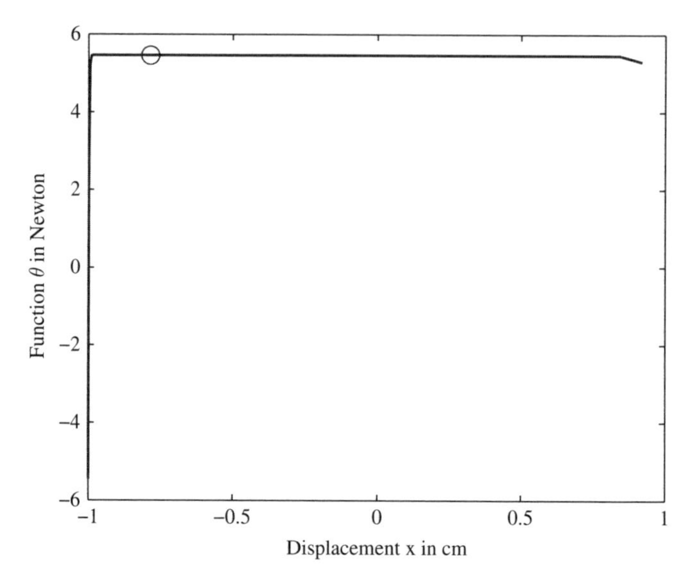

Figure 5.16 Function $\theta(x)$. The marker corresponds to the point whose abscissa is x_{*1}.

Step 3

The function $\theta(x)$ is computed using Equations (5.123) and (5.124). This function is shown in Figure 5.16. It can be seen that the zero $x_* = -0.9985\,\text{cm}$ of the function θ is close to -1.

Step 4

The value of the parameter a is determined from Equation (5.126). Therefore

$$a = 3545.5\,\text{N/cm}$$

Step 5

The value $x_{*1} = -0.7862\,\text{cm}$ is chosen to determine the remaining values of ρ and κ_w. Any value of $x_{*1} > x_*$ will lead to the same values for these parameters. Therefore

$$\rho = 649.18\,\text{cm}^{-1} \qquad \text{and} \qquad \kappa_w = 5.4615\,\text{N}$$

The theoretical values of ρ and κ_w are $\rho = 600\,\text{cm}^{-1}$ and $\kappa_w = 5.46\,\text{N}$, respectively. Note that there is a relative error of 8.2% in the value of ρ, while the relative error in the parameter κ_w is 0.03%. The following analysis explains this observation.

Note that, in a practical case, there may be some uncertainty on the parameter a, which is the main slope of the function θ in the region where this function is almost vertical (see Figure 5.16). Next it is shown that this possible uncertainty on the parameter a has little influence on the Dahl model. Indeed, Equation (5.125) shows that

$$\kappa_w = \frac{\left[\dfrac{d\theta(x)}{dx}\right]_{x_{*1}}}{\rho} + \theta\,(x_{*1}) \tag{5.135}$$

Since $\rho X_{\max}$ has a large value with respect to unity, and since the derivative

$$\left[\frac{d\theta(x)}{dx}\right]_{x_{*1}}$$

is small due to the fact that x_{*1} has been chosen in the region where the function θ is almost constant (see Figure 5.16), then

$$\frac{\left[\dfrac{d\theta(x)}{dx}\right]_{x_{*1}}}{\rho} \approx 0 \tag{5.136}$$

Combining Equations (5.135) and (5.136), it follows that $\kappa_w \approx \theta\,(x_{*1})$, which means that κ_w is insensitive to the possible uncertainty on the parameter a.

Alternatively, Equations (5.126) and (5.128) show that the slope of the function $\theta(x)$ at the point x_* is $a = \kappa_w \rho$. Since the parameter κ_w is insensitive to the uncertainty on a, the relative error on a is equal to the relative error on ρ. However, a precise determination of the parameter ρ is not relevant as the Dahl model is close to the Coulomb model for dry friction when the quantity $\rho X_{\max}$ is large. This fact is illustrated by considering that the obtained value of the parameter a has an uncertainty between −50% and +100%; that is the true value of a is $a/2$ and $2a$, respectively, which corresponds to $a = 1750\,\mathrm{N/cm}$ and $a = 7000\,\mathrm{N/cm}$. Then the corresponding values of the Dahl model parameters are obtained as

$$\kappa_w = 5.4615\,\mathrm{N}, \qquad \rho = 320.4\,\mathrm{cm}^{-1}$$

and

$$\kappa_w = 5.4615\,\mathrm{N}, \qquad \rho = 1281.7\,\mathrm{cm}^{-1}$$

respectively.

Figure 5.17 presents the responses of the viscous + Dahl model for the three values of ρ. These responses are similar, which means that the dynamics of the viscous + Dahl model is not sensitive to the value of ρ when the quantity $\rho X_{\max}$ is large with respect to unity.

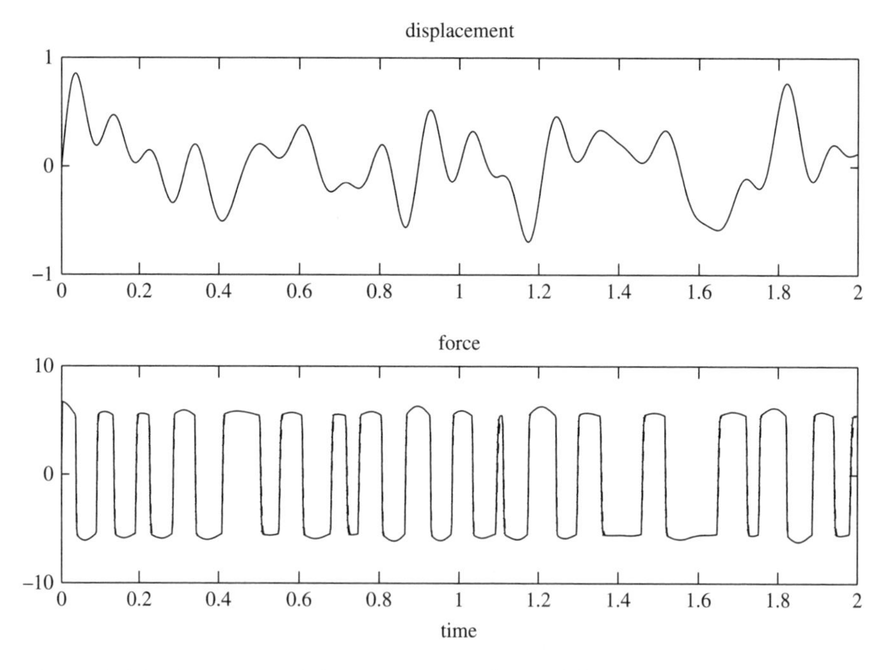

Figure 5.17 Response of the viscous + Dahl model to a random input signal with a range of frequencies that covers the interval [0,10 Hz]: solid, $\rho = 600\,\mathrm{cm}^{-1}$; dashed, $\rho = 320.4\,\mathrm{cm}^{-1}$; dotted, $\rho = 1281.7\mathrm{cm}^{-1}$.

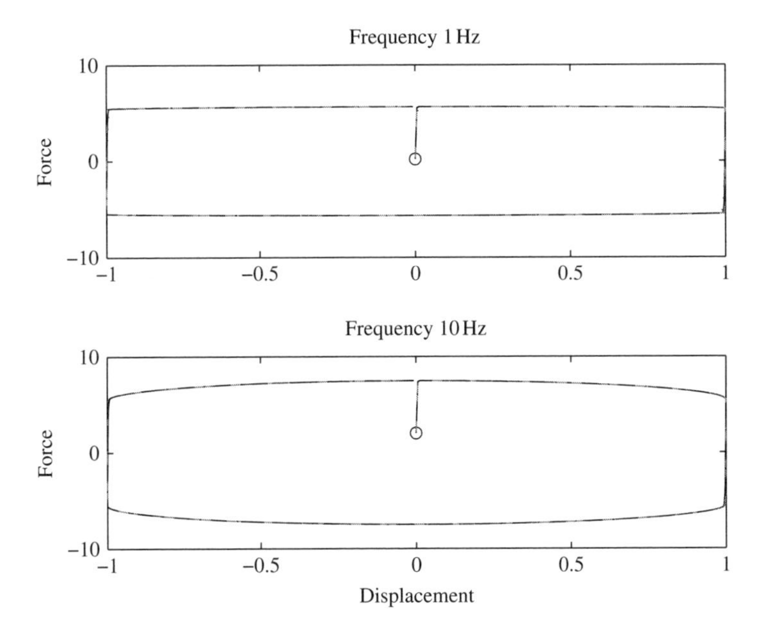

Figure 5.18 Force in N versus displacement in cm: solid, viscous + Bouc–Wen; dotted, viscous + Dahl with $\rho = 320.4\,\text{cm}^{-1}$; dashed, viscous + Dahl with $\rho = 1281.7\,\text{cm}^{-1}$; dotted-dashed, viscous + Dahl with $\rho = 649.18\,\text{cm}^{-1}$.

Figure 5.18 gives the responses of the viscous + Bouc-Wen and viscous + Dahl models to the input displacement signal $x(t) = \sin(2\pi ft)$. The upper curves correspond to the frequency $f = 1\,\text{Hz}$ and the lower curves correspond to the frequency $f = 10\,\text{Hz}$. The response of the viscous + Bouc–Wen model corresponds to the MR damper parameters (5.100), while the response of the viscous + Dahl model corresponds to the three sets of parameters: $\kappa_{xa} = 0.0320\,\text{N s/cm}$, $\kappa_w = 5.4615$ with $\rho = 320.4\,\text{cm}^{-1}$, $\rho = 649.18\,\text{cm}^{-1}$ and $\rho = 1281.7\,\text{cm}^{-1}$. The marker corresponds to the instant time $t = 0$. A good agreement is observed.

The corresponding force versus velocity plot is given in Figure 5.19. When the identified value of the parameter ρ is close to its nominal one ($\rho = 600\,\text{cm}^{-1}$), a good match is observed between the responses. However, the force/velocity plot shows more sensitivity to an uncertainty on the identified value of ρ than the force/displacement plot.

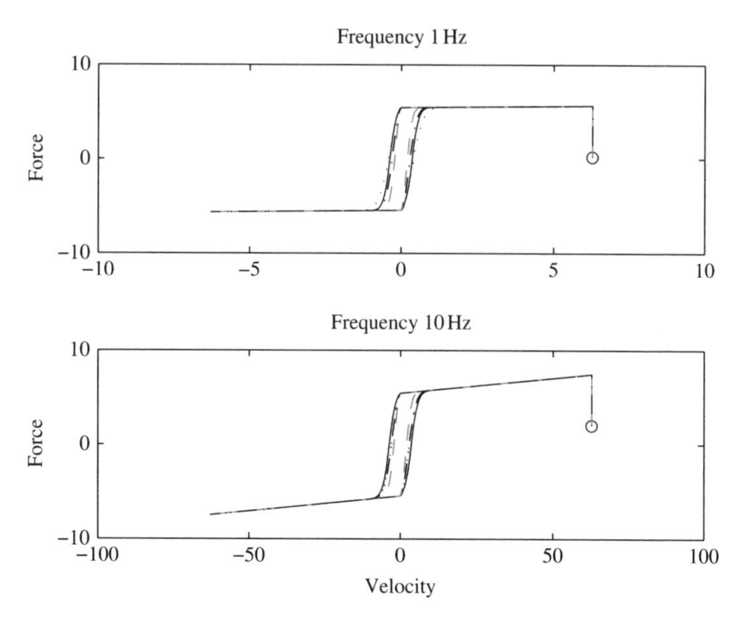

Figure 5.19 Force in N versus velocity in cm/s: solid, viscous + Bouc–Wen; dotted, viscous + Dahl with $\rho = 320.4\,\mathrm{cm}^{-1}$; dashed, viscous + Dahl with $\rho = 1281.7\,\mathrm{cm}^{-1}$; dotted-dashed, viscous + Dahl with $\rho = 649.18\,\mathrm{cm}^{-1}$.

5.4 CONCLUSION

This chapter has presented an identification method for the Bouc–Wen model that uses the analytical description of the limit cycle for the Bouc–Wen model. The method consists in exciting the hysteretic system with two input signals that differ by a constant and use the obtained limit cycles to derive the parameters of the Bouc–Wen model. This technique provides the exact values of the parameters in the absence of disturbances and proves to be robust with respect to a class of perturbations of practical relevance. The implementation of the identification methodology has been illustrated by means of numerical simulation. As an application of the results obtained in this and previous chapters, a case study of MR dampers has been analysed. It has been shown that the use of the frictional Dahl model is sufficient to describe these devices.

6

Control of a System with a Bouc–Wen Hysteresis

6.1 INTRODUCTION AND PROBLEM STATEMENT

In this chapter, the second-order mechanical system described by

$$m\ddot{x} + c\dot{x} + \Phi(x)(t) = u(t) \tag{6.1}$$

is considered with initial conditions $x(0)$, $\dot{x}(0)$ and excited by a control input force $u(t)$. The restoring force Φ is assumed to be described by the normalized Bouc–Wen model

$$\Phi(x)(t) = \kappa_x x(t) + \kappa_w w(t) \tag{6.2}$$

$$\dot{w}(t) = \rho\left(\dot{x}(t) - \sigma|\dot{x}(t)|\,|w(t)|^{n-1}w(t) + (\sigma - 1)\dot{x}(t)|w(t)|^n\right) \tag{6.3}$$

with an initial condition $w(0)$. The parameters

$$n \geq 1,\ \ \rho > 0,\ \ \sigma \geq \frac{1}{2},\ \ \kappa_x > 0,\ \ \kappa_w > 0,\ \ m > 0 \text{ and } c \geq 0$$

are unknown. The displacement $x(t)$ and velocity $\dot{x}(t)$ are available through measurements, but the signal $w(t)$ is not. The control input $u(t)$ is to be designed.

Systems with Hysteresis: Analysis, Identification and Control using the Bouc–Wen Model
F. Ikhouane and J. Rodellar

Let $y_r(t)$ be a (known) smooth and bounded reference signal whose (known) smooth and bounded derivatives are such that

$$\lim_{t\to\infty} y_r(t) = \lim_{t\to\infty} \dot{y}_r(t) = \lim_{t\to\infty} \ddot{y}_r(t) = \lim_{t\to\infty} y_r^{(3)}(t) = 0$$

exponentially. This means that there exists some constants $a > 0$ and $b > 0$ such that

$$\left|y_r^{(i)}(t)\right| \leq a\mathrm{e}^{-bt} \qquad \text{for } t \geq 0 \text{ and } \; i = 0, 1, 2, 3$$

The control objective is globally and asymptotically to regulate the displacement $x(t)$ and velocity $\dot{x}(t)$ to the reference signals $y_r(t)$ and $\dot{y}_r(t)$ preserving the global boundedness of all the closed-loop signals, that is $x(t)$, $\dot{x}(t)$, $w(t)$ and $u(t)$. Some information on the unknown parameters is assumed.

Assumption 3. *The unknown parameters lie in known intervals; that is* $m \in [m_{\min}, m_{\max}]$ *with* $m_{\min} > 0$, $c \in [0, c_{\max}]$, $\kappa_x \in \left(0, \kappa_{x_{\max}}\right]$, $\kappa_w \in \left(0, \kappa_{w_{\max}}\right]$, $\sigma \in [1/2, \sigma_{\max}]$ *and* $\rho \in (0, \rho_{\max}]$.

Note that the unknown structure parameter $n \geq 1$ is not required to lie in a known interval.

The problem of controlling the system (6.1)–(6.3) has been treated in Reference [53] using adaptive control to obtain a global boundedness result and a region of ultimate boundedness as small as desired. The control law of this chapter improves the results of Reference [53] by showing that, under PID control, the displacement and velocity errors tend to zero (not to a neighbourhood of the origin) as time increases, as frequently demanded in applications.

This problem lies within the general context of the regulation problem of nonlinear systems in the presence of uncertain dynamics and uncertainties in the parameters. Much attention has been devoted to this kind of problem in the current literature. In Reference [140] the nonlinear system is split into two interconnected parts: an unmeasured zero dynamics block, which is assumed to be exponentially stable, and a triangular block with uncertain control gain. Under reasonable assumptions, an integral action is added to a sliding mode control law. The sliding mode part drives the tracking error to a neighbourhood of the origin, and the integral action takes it asymptotically to zero. The stability results are semi-global.

In Reference [141], a nonlinear system with a relative degree of one is studied under the assumption that the reference signal and the output disturbances are produced by a Poisson stable system. Necessary conditions for output regulation are obtained using topological concepts derived from the notion of Poisson stability. Under the additional assumption that perfect tracking can be achieved when the feedforward input signals belong to the set of solutions of a suitable differential equation, a systematic method for the design of a controller that solves the problem of output regulation is obtained. Asymptotic regulation is achieved provided that the limit set of initial conditions of the zero dynamics augmented by the dynamic of the reference is locally exponentially attractive. The stability results are semi-global.

In Reference [142], the class of nonlinear systems under consideration is composed of unmeasured, input-to-state stable (ISS) zero dynamics coupled with a block strict-feedback part that is driven by an unknown function that may depend on measured outputs and the zero dynamics. Under additional technical assumptions and using the backstepping technique, a control law design procedure that insures global stability and asymptotic regulation is derived.

Note that the results of the references above do not allow the proposed control objective to be achieved, either because of its nature (semi-global stability in References [140] and [141] instead of the global stability sought) or because the system under study does not fulfill the required conditions (the present system does not satisfy condition A2 in Reference [142], and also the Bouc–Wen model is not ISS, as demanded by Reference [142]). Some peculiarities of the Bouc–Wen model are exploited in this chapter to show that global stability and asymptotic regulation can be obtained under PID control.

6.2 CONTROL DESIGN AND STABILITY ANALYSIS

In this section, it is shown that a PID control achieves the regulation of the displacement x and velocity $\dot{x}$ to the exponentially decaying reference signals y_r and $\dot{y}_r$ respectively. To this end, the following variables are introduced:

$$x_1(t) = x(t) - y_r(t)$$

$$x_2(t) = \dot{x}(t) - \dot{y}_r(t)$$

$$x_0(t) = \int_0^t x_1(\tau)\mathrm{d}\tau$$

and the PID controller is chosen as a control law:

$$u(t) = -k_0 x_0(t) - k_1 x_1(t) - k_2 x_2(t) \tag{6.4}$$

where the k_i values are design parameters.

The main result of this chapter is summarized in the following theorem.

Theorem 6. *Consider the closed loop formed by the system* (6.1)–(6.3) *and the control law* (6.4). *Define the following constants*:

$$k_{2_{\min}} = \sqrt{2m_{\max}\left(\sigma_{\max}\rho_{\max}\kappa_{w_{\max}} + \kappa_{x_{\max}} + k_1\right)} \tag{6.5}$$

$$e_1 = \frac{\left(c_{\max} + k_2\right)^3}{m_{\min}^2}$$

$$e_2 = \frac{k_1^2}{m_{\max}^2}\left(k_2^2 - k_{2_{\min}}^2\right)$$

$$k_{0_{\max}} = \min\left(\frac{k_1 k_2}{m_{\max}}, -e_1 + \sqrt{e_1^2 + e_2}\right) \tag{6.6}$$

and choose the design gains k_0, k_1 and k_2 in the following way:

(*a*) *Take any positive value for k_1.*
(*b*) *Then, choose k_2 such that $k_2 > k_{2_{\min}}$.*
(*c*) *Finally, take $0 < k_0 < k_{0_{\max}}$.*

This gives the following two parts:

1. *All the closed-loop signals x_0, x_1, x_2, w and the control u are globally bounded.*
2. $\lim_{t\to\infty} x(t) = 0$ *and* $\lim_{t\to\infty} \dot{x}(t) = 0$.

Proof. Part 1 of Theorem 6 is first proved. The closed loop is described by the four-state system:

$$\dot{x}_0 = x_1 \tag{6.7}$$

$$\dot{x}_1 = x_2 \tag{6.8}$$

$$\dot{x}_2 = -m^{-1}\left[(c+k_2)\,x_2 + (\kappa_x + k_1)\,x_1 + k_0 x_0)\right] - m^{-1}\left(\kappa_w w + m\ddot{y}_r + c\dot{y}_r + \kappa_x y_r\right) \tag{6.9}$$

$$\dot{w} = \rho\left(x_2 + \dot{y}_r - \sigma|x_2 + \dot{y}_r|\;|w|^{n-1} w + (\sigma - 1)(x_2 + \dot{y}_r)\,|w|^n\right) \tag{6.10}$$

Equation (6.9) can be written as

$$m x_0^{(3)} + (c+k_2)\,\ddot{x}_0 + (\kappa_x + k_1)\,\dot{x}_0 + k_0 x_0 = -\kappa_w w - m\ddot{y}_r - c\dot{y}_r - \kappa_x y_r \tag{6.11}$$

The constants k_0, k_1 and k_2 are chosen so that the linear system in the left-hand part of Equation (6.11) is exponentially stable. Using a Routh argument, it is immediately seen that this happens if the following conditions are satisfied:

$$c + k_2 > 0 \tag{6.12}$$

$$0 < k_0 < \frac{(c+k_2)(k_1+\kappa_x)}{m} \tag{6.13}$$

Since the parameters c, κ_x and m are unknown but bounded, as demanded by Assumption 3, sufficient conditions to satisfy (6.12) and (6.13) are given by

$$k_1 > 0, \quad k_2 > 0 \quad \text{and} \quad 0 < k_0 < \frac{k_1 k_2}{m_{\max}}$$

At this point, a result obtained in Chapter 2 is used.

Theorem 7. *Consider the nonlinear differential equation* (6.3) *as a continuous time system whose input and output are* $\dot{x}(t)$ *and* $w(t)$*, respectively. Then, the output signal* $w(t)$ *is bounded for any initial condition* $w(0)$ *and any continuous function* $\dot{x}(t)$ *(be it bounded or not). Furthermore,*

$$|w(t)| \leq \max(1, |w(0)|) \qquad \textit{for all } t \geq 0$$

A by-product of Theorem 7 is the existence and uniqueness of the solutions of the system (6.7)–(6.10) over the time interval $[0, +\infty)$. Also, due to the boundedness of $w(t)$, Equation (6.11) may be seen as an exponentially stable linear system excited by a bounded input. This implies that the signals $x_0(t)$, $x_1(t)$ and $x_2(t)$ are bounded, which implies the boundedness of the control signal $u(t)$ (see Equation (6.4)).

Moving now to part 2 of Theorem 6, the key argument of the proof is to demonstrate that the velocity $x_2 \in L^1[0, \infty)$. This is done in Lemma 9 below. This fact, along with Equation (6.8), shows that x_1 has a finite limit (see Lemma 13 in the Appendix). This limit has to be zero, otherwise the state x_0 would go to infinity, which contradicts the above proved boundedness. Since x_2 is bounded and $\dot{x}_2$ is bounded (by Equation (6.9)) and $x_2 \in L^1$, then by Barbalat's lemma (Lemma 14 in the Appendix) it follows that

$$\lim_{t\to\infty} x_2(t) = 0$$

Therefore, it has been shown that a PID control with appropriately chosen gains ensures that

$$\lim_{t\to\infty} [x(t) - y_r(t)] = \lim_{t\to\infty} x(t) = 0$$

and

$$\lim_{t\to\infty} [\dot{x}(t) - \dot{y}_r(t)] = \lim_{t\to\infty} \dot{x}(t) = 0$$

along with the boundedness of all the closed-loop signals, that is x_0, x (and x_1), $\dot{x}$ (and x_2), w and the control u.

Lemma 9. *The velocity x_2 belongs to $L^1[0, \infty)$.*

Proof. Two cases are discussed:

P_1: $|w(t)| > 1$ for all $t \geq 0$.
P_2: There exists some $t_0 < \infty$ such that $|w(t_0)| \leq 1$.

The case P_1 is treated first. From Chapter 2,

$$w(t) = \frac{z(t)}{z_0}$$

which implies that

$$w^2(t) = 2\frac{V(t)}{z_0^2} \qquad \text{where} \quad V(t) = \frac{z^2(t)}{2}$$

On the other hand, it has been demonstrated in Chapter 2 that the time function $V(t)$ is nonincreasing. This fact implies that the function $w(t)^2$ is nonincreasing. Since it is bounded, it goes to a limit $w_\infty^2 \geq 1$. Consider the case where $w(0) > 0$ (the analysis is similar in the case $w(0) < 0$). Then, by the continuity of w,

$$w(t) \geq w_\infty \geq 1 \qquad \text{for all } t \geq 0$$

Take $\varepsilon > 0$; then some $t_1 < \infty$ exists such that

$$w_\infty^n \leq w(t)^n \leq w_\infty^n + \varepsilon \qquad \text{for all } t \geq t_1 \tag{6.14}$$

Multiplying by $x_2 + \dot{y}_r$ and integrating both parts of (6.14), the following is obtained for any $T \geq 0$:

$$\begin{aligned}
& w_\infty^n \int_{t_1}^{t_1+T} [x_2(t) + \dot{y}_r(t)]\,\mathrm{d}t - \varepsilon \int_{t_1}^{t_1+T} |x_2(t) + \dot{y}_r(t)|\,\mathrm{d}t \\
& \leq \int_{t_1}^{t_1+T} [x_2(t) + \dot{y}_r(t)]\, w(t)^n \mathrm{d}t \\
& \leq w_\infty^n \int_{t_1}^{t_1+T} [x_2(t) + \dot{y}_r(t)]\,\mathrm{d}t + \varepsilon \int_{t_1}^{t_1+T} |x_2(t) + \dot{y}_r(t)|\,\mathrm{d}t
\end{aligned} \tag{6.15}$$

On the other hand, from Equations (6.10) and (6.14) the following is obtained for all $t \geq t_1$:

$$\begin{aligned}
\sigma\rho\, |x_2 + \dot{y}_r|\, w_\infty^n & \leq \sigma\rho\, |x_2 + \dot{y}_r|\, w^n \\
& = -\dot{w} + \rho\,(x_2 + \dot{y}_r) + \rho(\sigma - 1)\,(x_2 + \dot{y}_r)\, w^n
\end{aligned} \tag{6.16}$$

Integrating both parts of inequality (6.16) and using Equation (6.15), it follows that

$$\begin{aligned}
\int_{t_1}^{t_1+T} |x_2(t) + \dot{y}_r(t)|\,\mathrm{d}t \leq & -\frac{1}{\sigma\rho w_\infty^n}\,[w(t_1 + T) - w(t_1)] \\
& + \left(\frac{1}{\sigma w_\infty^n} - \frac{\sigma - 1}{\sigma}\right) [x_1(t_1 + T) - x_1(t_1)
\end{aligned}$$

$$+y_r(t_1+T)-y_r(t_1)]$$
$$+\frac{|\sigma-1|\varepsilon}{\sigma w_\infty^n}\int_{t_1}^{t_1+T}|x_2(t)+\dot{y}_r(t)|\,\mathrm{d}t \qquad (6.17)$$

If $\sigma=1$, it follows from Equation (6.17) and the boundedness of the signals x_1 and y_r that $x_2+\dot{y}_r\in L^1$ as T is arbitrary. If $\sigma\neq 1$, choosing

$$\varepsilon=\frac{\sigma w_\infty^n}{2|\sigma-1|}$$

in Equation (6.17) shows that $x_2+\dot{y}_r\in L^1$. Since it has been assumed that $\dot{y}_r$ goes exponentially to zero, then $\dot{y}_r\in L^1$. Then, it follows that $x_2\in L^1$.

Now turn to the case P_2. Taking the derivative of Equation (6.11) gives

$$\lambda_3 x_1^{(3)}+\lambda_2\ddot{x}_1+\lambda_1\dot{x}_1+\lambda_0 x_1=-\dot{w}-\kappa \qquad (6.18)$$

where

$$\begin{aligned}\lambda_3&=\frac{m}{\kappa_w},\ \lambda_2=\frac{c+k_2}{\kappa_w},\ \lambda_1=\frac{\kappa_x+k_1}{\kappa_w},\ \lambda_0=\frac{k_0}{\kappa_w},\\ \kappa(t)&=\frac{m}{\kappa_w}y_r^{(3)}(t)+\frac{c}{\kappa_w}\ddot{y}_r(t)+\frac{\kappa_x}{\kappa_w}\dot{y}_r(t)\end{aligned} \qquad (6.19)$$

By Assumption P_2, $|w(t_0)|\leq 1$. Using Theorem 7, it follows that $|w(t)|\leq 1$ for all $t\geq t_0$. From Equation (6.10), for all $t\geq t_0$,

$$\begin{aligned}0\leq\dot{w}\,(x_2+\dot{y}_r)&=\rho\,(1-|w|^n)\,(x_2+\dot{y}_r)^2\\ &\leq\rho\,(x_2+\dot{y}_r)^2 \qquad \text{for } (x_2+\dot{y}_r)\,w\geq 0\\ \rho\,(x_2+\dot{y}_r)^2\leq\dot{w}\,(x_2+\dot{y}_r)&=\rho\,(1+(2\sigma-1)|w|^n)\,(x_2+\dot{y}_r)^2\\ &\leq 2\sigma\rho\,(x_2+\dot{y}_r)^2 \qquad \text{for } (x_2+\dot{y}_r)\,w\leq 0\end{aligned}$$

where the fact has been used that $\sigma\geq\frac{1}{2}$. Thus, in all cases,

$$0\leq\dot{w}\,(x_2+\dot{y}_r)\leq\varrho\,(x_2+\dot{y}_r)^2 \qquad (6.20)$$

where $\varrho=2\sigma\rho$.

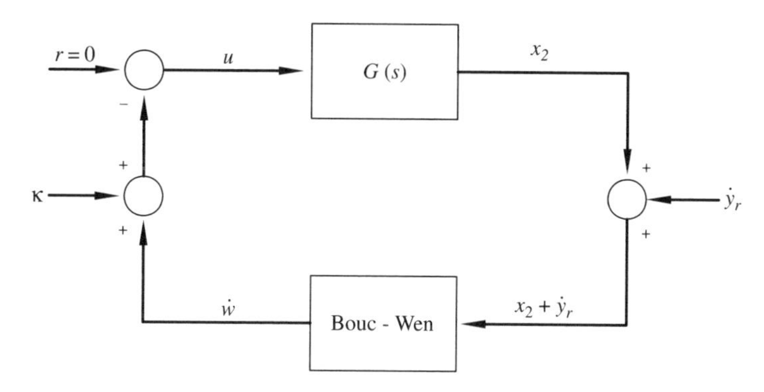

Figure 6.1 Equivalent description of Equation (6.18).

Note that Equation (6.18) can be viewed as a feedback connection as in Figure 6.1, where the reference is $r = 0$, the control signal is $u = -(\dot{w} + \kappa)$ and

$$G(s) = \frac{s}{\lambda_3 s^3 + \lambda_2 s^2 + \lambda_1 s + \lambda_0}$$

By Equation (6.20), the Bouc–Wen nonlinearity in Figure 6.1, with input $x_2 + \dot{y}_r$ and output $\dot{w}$, belongs to the sector $[0, \varrho]$ (see Section A.3.3 in the Appendix).

The idea is to use this fact to prove that the feedback connection is such that the state x_2 goes exponentially to zero using Theorem 13 (in the Appendix) once it has been proved that the transfer function $1 + \varrho G(s)$ is strictly positive real. It is to be noted that the Bouc–Wen nonlinearity is not memoryless in the present case. However, w is bounded by Theorem 7, x_2 has been shown to be bounded above and the variable $\dot{x}_2$ is bounded as all the quantities in the right-hand side of Equation (6.9) are bounded. Then, it can easily be checked that the stability proof of the feedback connection of Figure 6.1 is a small variation of Theorem 13 taking into account the boundedness of the states x_2 and $\dot{x}_2$ along with the exponential decay of the reference signal and its derivatives to zero. Thus, if it can be proved that the transfer function $1 + \varrho G(s)$ is strictly positive real, it will follow from the proof of Theorem 13 that the states x_2 and $\dot{x}_2$ go exponentially to zero (the remaining state w of the closed loop is bounded but does not necessarily go to zero).

Now, it remains to determine the conditions on the gains k_0, k_1 and k_2 so that the transfer function $1 + \varrho G(s)$ is strictly positive real. In the following, the simplified version of Theorem 13 given in

Theorem 14 (see the Appendix) is used. It should be checked whether $\mathrm{Re}\left[1+\varrho G(j\omega)\right] > 0$ for all $\omega \in [-\infty, \infty]$; this is equivalent to finding conditions under which

$$p(\eta) = \eta^3 + a_1\eta^2 + a_2\eta + a_3 > 0 \tag{6.21}$$

for all $\eta = \omega^2 \geq 0$, where

$$a_1 = \frac{-\lambda_3\varrho + \lambda_2^2 - 2\lambda_1\lambda_3}{\lambda_3^2}, \qquad a_2 = \frac{\varrho\lambda_1 - 2\lambda_0\lambda_2 + \lambda_1^2}{\lambda_3^2},$$
$$a_3 = \frac{\lambda_0^2}{\lambda_3^2} > 0 \tag{6.22}$$

It is clear from Equation (6.21) that having $p(\eta) > 0$ for all $\eta \geq 0$ is equivalent to the stability of the polynomial p. A Routh argument gives the following conditions of stability:

$$\lambda_2 > \sqrt{\varrho\lambda_3 + 2\lambda_1\lambda_3} \tag{6.23}$$

$$\lambda_0^2 + f_1\lambda_0 - f_2 < 0 \tag{6.24}$$

where

$$f_1 = \frac{2\lambda_2\left(-\lambda_3\varrho + \lambda_2^2 - 2\lambda_1\lambda_3\right)}{\lambda_3^2} \tag{6.25}$$

$$f_2 = \frac{1}{\lambda_3^2}\left(-\lambda_3\varrho + \lambda_2^2 - 2\lambda_1\lambda_3\right)\left(\varrho\lambda_1 + \lambda_1^2\right) \tag{6.26}$$

Thus, the gains k_0, k_1 and k_2 will be chosen as follows. For k_1 any positive value is taken. Then, $k_2 > 0$ is chosen such that the inequality (6.23) holds. Since the constants that appear in this equation are unknown, a sufficient condition to get (6.23) is by choosing

$$k_2 > \sqrt{2m_{\max}\left(\sigma_{\max}\rho_{\max}\kappa_{w_{\max}} + \kappa_{x_{\max}} + k_1\right)} = k_{2_{\min}} \tag{6.27}$$

Since the parameters f_1 and f_2 are uncertain, a sufficient condition to have (6.24) is by choosing

$$0 < k_0 < -e_1 + \sqrt{e_1^2 + e_2} \tag{6.28}$$

where

$$e_1 = \frac{(c_{\max} + k_2)^3}{m^2_{\min}}$$

$$e_2 = \frac{k_1^2}{m^2_{\max}} \left(k_2^2 - k^2_{2_{\min}}\right) \tag{6.29}$$

Note that $e_2 > 0$ by Equation (6.27). Recall that

$$0 < k_0 < \frac{k_1 k_2}{m_{\max}}$$

to comply with the stability condition in (6.13). Thus, k_0 needs to be chosen so that

$$0 < k_0 < \min\left(\frac{k_1 k_2}{m_{\max}}, -e_1 + \sqrt{e_1^2 + e_2}\right) = k_{0_{\max}} \tag{6.30}$$

It has therefore been proved that the gains k_0, k_1 and k_2 can be chosen so that the transfer function $1 + \varrho G(s)$ is strictly positive real. This implies that the exponential decay of x_2 goes to zero and thus that $x_2 \in L^1$.

It has thus been proved that, in all cases P_1 and P_2, $x_2 \in L^1$.

6.3 NUMERICAL SIMULATION

The following numerical simulation illustrates the effectiveness of the control scheme. The following values for the parameters are taken:

$$m = 1, \quad c = 1, \quad \kappa_x = 1, \quad \kappa_w = 1, \quad \sigma = 1, \quad \rho = 1, \quad n = 1.5$$

These values are supposed to be unknown, and it is known instead that $m \in [m_{\min} = 0.5, m_{\max} = 2]$ and for any parameter $p \in \{c, \kappa_x, \kappa_w, \rho, \sigma\}$, it is considered that $p_{\max} = 2$.

The PID design parameters are chosen following Theorem 6:

1. First $k_1 = 1000$ is chosen which gives $k_{2_{\min}} = 63.5610$.
2. Then, $k_2 = k_{2_{\min}} + 5$ is taken, which gives $k_{0_{\max}} = 58.7614$.
3. Finally, $k_0 = 0.9 k_{0_{\max}}$ is taken.

For the reference signal, y_r is chosen as the output of the second-order linear system

$$\frac{\omega_0^2}{s^2 + 2\xi\omega_0 s + \omega_0^2}$$

with $\xi = 0.7$, $\omega_0 = 1$ and zero input; that is the linear system is driven only by the nonzero initial conditions $y_r(0) = x(0)$ and $\dot{y}_r(0) = \dot{x}(0)$.

Figure 6.2 shows the behaviour of the closed loop starting from the initial conditions: $x(0) = 1$, $\dot{x}(0) = 1$ and $w(0) = 0$. The regulation is achieved both for the displacement output and for the velocity. During the transient, it is observed that the errors have a small magnitude compared with the size of the initial conditions, with a seemingly reasonable control effort.

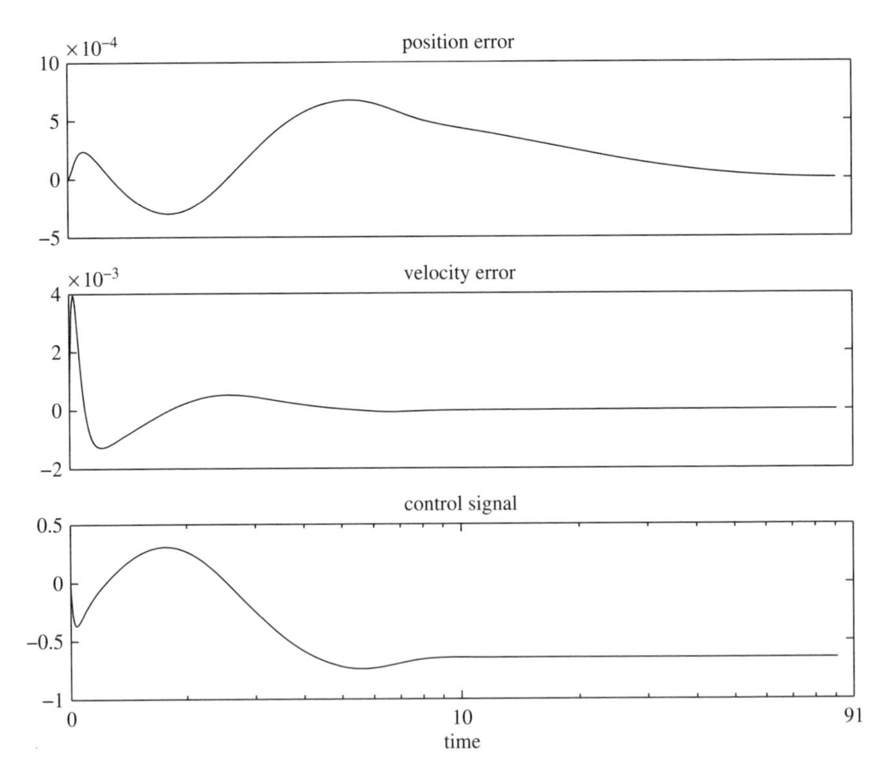

Figure 6.2 Closed-loop signals.

6.4 CONCLUSION

This chapter has focused on the problem of regulating the displacement and velocity of a second-order system that includes a dynamic hysteresis described by the Bouc–Wen model. It has been shown that the boundedness of all closed loop signals as well as regulation of the displacement and velocity can be achieved by a simple PID controller. The closed-loop stability analysis exploits the peculiarities of the Bouc–Wen model dynamics to derive the desired results analytically. A numerical simulation has shown the effectiveness of the PID control.

Appendix

Mathematical Background

This appendix presents a brief summary of some results that are used throughout the book. It is based on References [143], [144] and [145].

A.1 EXISTENCE AND UNIQUENESS OF SOLUTIONS

Consider the differential equation

$$\dot{x} = f(t, x), \qquad x(t_0) = x_0 \tag{A.1}$$

where $x(t) \in \mathbb{R}^n$ is the state vector and n is a positive integer. The function f is defined from $\mathbb{R} \times \mathbb{R}^n$ to $\mathbb{R}^n$. A solution of the differential equation (A.1) on the time interval $[t_0, t_1]$ is a continuous function $x : [t_0, t_1] \to \mathbb{R}^n$ such that $\dot{x}(t)$ is defined and $\dot{x}(t) = f(t, x(t))$ for all $t \in [t_0, t_1]$. If f is continuous in (t, x) then the solution $x(t)$ will be continuously differentiable. It is assumed that f is continuous in x but only piecewise continuous in t. In this case, the solution can only be piecewise continuously differentiable.

The existence and uniqueness of solutions of the differential equation (A.1) is guaranteed when the function f is locally Lipschitz in x. This means that positive constants L and r exist such that

$$\|f(t, x) - f(t, y)\| \leq L\|x - y\|$$

for all $x, y \in B = \{x \in \mathbb{R}^n / ||x - x_0|| \leq r\}$ and for all $t \in [t_0, t_1]$.

Systems with Hysteresis: Analysis, Identification and Control using the Bouc–Wen Model
F. Ikhouane and J. Rodellar

The following theorem states a sufficient condition for the local existence and uniqueness of solutions of the differential equation (A.1).

Theorem 8. *Assume that the function f is piecewise continuous in t and locally Lipschitz in x. Then, a positive real δ exists such that the differential equation (A.1) has a unique solution on the time interval $[t_0, t_0 + \delta]$.*

Note that it is not necessary to assume the continuity in x of the function f as this is implied by the Lipschitz condition. The global existence and uniqueness of solutions is guaranteed when the Lipschitz condition holds globally.

Theorem 9. *Assume that the function $f(t, x)$ is piecewise continuous in t and globally Lipschitz in x; that is a positive constant L exists such that*

$$\|f(t, x) - f(t, y)\| \leq L\|x - y\|$$

for all $x, y \in \mathbb{R}^n$ and for all $t \in [t_0, t_1]$. Then, the differential equation (A.1) has a unique solution on the time interval $[t_0, t_1]$.

In view of the conservative nature of the global Lipschitz condition, it would be useful to have a global existence and uniqueness theorem that requires the function f to be only locally Lipschitz. The next theorem achieves that at the expense of having to know more about the solution of the system.

Theorem 10. *Assume that the function $f(t, x)$ is piecewise continuous in t and locally Lipschitz in x for all $t \geq t_0$ and all x in a domain $D \subset \mathbb{R}^n$. Let W be a compact subset of D, $x_0 \in W$, and suppose it is known that every solution of*

$$\dot{x} = f(t, x), \qquad x(t_0) = x_0 \tag{A.2}$$

lies entirely in W. There is then a unique solution that is defined for all $t \geq t_0$.

A.2 CONCEPTS OF STABILITY

Consider now the differential equation (A.1), where $f: [0, +\infty) \times D \to \mathbb{R}^n$ is piecewise continuous in t and locally Lipschitz in x on $[0, +\infty) \times D$. The set $D \subset \mathbb{R}^n$ is said to be a domain (that is an open connected set) that contains the origin $x = 0$. The origin is said to be an equilibrium point for (A.1) if $f(t, 0) = 0$ for all $t \geq 0$.

Assume that $x = 0$ is an equilibrium point for (A.1). It is said that $x = 0$ is locally uniformly stable if for each $\varepsilon > 0$, there is a $\delta = \delta(\varepsilon) > 0$ such that $||x(t_0)|| < \delta \Rightarrow ||x(t)|| < \varepsilon$ for all $t \geq t_0$. The following result gives a sufficient condition for the uniform stability of the equilibrium $x = 0$.

Theorem 11. *Let $x = 0$ be an equilibrium point for (A.1) and $D \subset \mathbb{R}^n$ be a domain containing $x = 0$. Let $V: [0, \infty) \times D \to \mathbb{R}$ be a continuously differentiable function such that*

$$W_1(x) \leq V(t, x) \leq W_2(x) \tag{A.3}$$

$$\frac{\partial V}{\partial t} + \frac{\partial V}{\partial x} f(t, x) \leq 0 \tag{A.4}$$

for all $t \geq 0$ and for all $x \in D$, where $W_1(x)$ and $W_2(x)$ are continuous positive definite functions[1] *on D. Then $x = 0$ is uniformly stable.*

V is called a Lyapunov function.

A special case of stability is exponential stability.

Definition 3. *The equilibrium point $x = 0$ of (A.1) is exponentially stable if positive constants c, k, and λ exist such that*

$$\|x(t)\| \leq k\|x(t_0)\| \mathrm{e}^{-\lambda(t-t_0)}, \qquad \forall \ \|x(t_0)\| < c \tag{A.5}$$

and globally exponentially stable if (A.5) is satisfied for any initial state $x(t_0)$.

In some kind of problems, it is not possible to have a Lyapunov function V whose derivative is always negative. In this case, it is not possible to guarantee the uniform stability of the system. However,

---

[1] The function $W(x)$ is said to be positive definite if (a) $W(x) \geq 0$ for all $x \in D$ and (b) $W(x) = 0$ is equivalent to $x = 0$. The function $W(x)$ is said to be positive semi-definite if condition (a) holds but (b) does not.

under some conditions, it is possible to show that the state $x(t)$ remains bounded.

Theorem 12. *Let $D \subset \mathbb{R}^n$ be a domain that contains the origin and*

$$V : [0, \infty) \times D \to \mathbb{R}$$

be a continuously differentiable function such that

$$\alpha_1(\|x\|) \leq V(t, x) \leq \alpha_2(\|x\|) \tag{A.6}$$

$$\frac{\partial V}{\partial t} + \frac{\partial V}{\partial x} f(t, x) \leq 0, \quad \forall \, \|x\| \geq \mu > 0 \tag{A.7}$$

for all $t \geq 0$ and for all $x \in D$, where $||x||$ is the Euclidean norm of the vector x and $\alpha_1(\cdot)$ and $\alpha_2(\cdot)$ are increasing continuous functions with $\alpha_1(0) = \alpha_2(0) = 0$.

Take $r > 0$ such that $B_r = \{x \in \mathbb{R}/||x|| < r\} \subset D$ and suppose that $\mu < \alpha_2^{-1}(\alpha_1(r))$. Then, for every initial state $x(t_0)$ satisfying $||x|| < \alpha_2^{-1}(\alpha_1(r))$, some $T \in [0, \infty]$ exists such that

$$\|x(t)\| \leq \alpha_1^{-1}(\alpha_2(x(t_0))), \qquad \forall \, t_0 \leq t \leq t_0 + T \tag{A.8}$$

$$\|x(t)\| \leq \alpha_1^{-1}(\alpha_2(\mu)), \qquad \forall \, t \geq t_0 + T \tag{A.9}$$

A.3 PASSIVITY AND ABSOLUTE STABILITY

A.3.1 Passivity in Mechanical Systems

Consider the mechanical system of Figure A.1 with mass m, damping $c(v) \geq 0$ and stiffness $k > 0$, where x is the displacement of the mass and v its velocity. This mass is subject to an external force f. The movement of the mass is described by the following differential equation:

$$m\dot{v} = -c(v)v - kx + f \tag{A.10}$$

Assume that the velocity is constant and that $k = 0$. Then, equation (A.10) reduces to

$$f - c(v)v = 0 \tag{A.11}$$

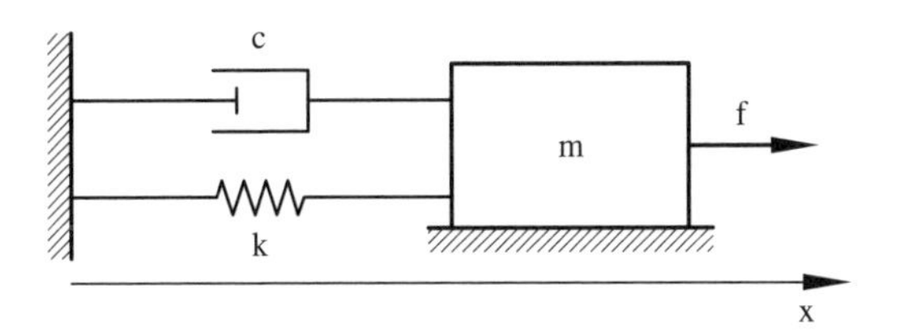

Figure A.1 Equivalent description of system (A.10).

The presence of a minus sign in the term $-c(v)v$ with $c(v) \geq 0$ means that the corresponding viscous friction force opposes the movement of the mass. Thus, the condition $c(v) \geq 0$ translates into a dissipation of the energy of this mass. If the mass is seen as a system whose input is the velocity v and output the force f, then the condition $c(v) \geq 0$ is equivalent to $vf \geq 0$. In systems theory, when the product of the input and the output of a nondynamical system is nonnegative, the system is said to be passive. In mechanical systems, passivity is related to energy dissipation.

Consider now the dynamical system (A.10). Multiplying both members of this equation by the velocity v gives

$$mv\dot{v} + c(v)v^2 + kvx = vf \tag{A.12}$$

Defining the function

$$V(t) = \frac{1}{2}mv^2 + \frac{1}{2}kx^2 \tag{A.13}$$

the following is obtained from Equations (A.12) and (A.13):

$$\dot{V}(t) + c(v)v^2 = vf \tag{A.14}$$

The condition $c(v) \geq 0$ that corresponds to energy dissipation translates into $v(t)f(t) \geq \dot{V}(t)$, which is the definition of passivity within systems theory.

More generally, consider a dynamical system with input u and output y given by the equations

$$\dot{h} = F(h, u) \tag{A.15}$$

$$y = G(h, u) \tag{A.16}$$

where $F: \mathbb{R}^n \times \mathbb{R} \to \mathbb{R}^n$ is locally Lipschitz, $G: \mathbb{R}^n \times \mathbb{R} \to \mathbb{R}$ is continuous and $F(0,0) = 0$, $G(0,0) = 0$. This system is said to be passive if a continuously differentiable positive semi-definite function $V(x)$ (called the storage function) exists such that

$$uy \geq \dot{V} = \frac{\partial V}{\partial x} F(x, u), \qquad \forall\, (x, u) \in \mathbb{R}^n \times \mathbb{R} \tag{A.17}$$

A.3.2 Positive Realness

Definition 4. *A rational function $H(s)$ of the complex variable s is said to be positive real (PR) if*

(*a*) *$H(s)$ is real for s real and*
(*b*) $\mathrm{Re}\,[H(s)] \geq 0$ *for all* $\mathrm{Re}\,[s] > 0$.

Strict positive realness is defined as follows.

Definition 5. *A rational function $H(s)$ is strictly positive real (SPR) if $H(s-\varepsilon)$ is PR for some $\varepsilon > 0$.*

The next lemma gives a necessary and sufficient condition for strict positive realness.

Lemma 10. *A proper rational function $H(s)$ is SPR if, and only if,*

$$H(s) \text{ is analytic in } \mathrm{Re}[s] \geq 0,$$

$$\mathrm{Re}\,[H(\mathrm{j}\omega)] > 0 \text{ for all } \omega \in (-\infty, \infty) \text{ and}$$

$$lim_{\omega^2 \to \infty} \omega^2\, \mathrm{Re}\,[H(\mathrm{j}\omega)] > 0, \text{ when } n^* = 1,$$

where n^* is the relative degree of $H(s)$, that is the number of poles of $H(s)$ minus the number of zeros of $H(S)$. If $H(s)$ is proper but not strictly proper (that is if $n^* = 0$), then conditions (a) and (b) are necessary and sufficient for $H(s)$ to be SPR.

From the definitions of rational PR and SPR functions above, it is clear that if $H(s)$ is PR, its phase shift for all frequencies lies in the interval $[-\pi/2, \pi/2]$. Hence, n^* can only be either 0 or 1 if $H(s)$ is the transfer function of a dynamical system that is causal.

Lemma 11. (*Kalman–Yakubovich–Popov*). *Let* $G(s) = C(sI - A)^{-1} B + D$ *be a transfer function, where* (A, B) *is controllable and* (A, C) *is observable. Then,* $G(s)$ *is strictly positive real if and only if there exist matrices* $P = P^T > 0$, L *and* W, *and a positive constant* ε *such that*

$$\begin{aligned} PA + A^T P &= -L^T L - \varepsilon P \\ PB &= C^T - L^T W \\ W^T W &= D + D^T \end{aligned} \tag{A.18}$$

A.3.3 Sector Functions

Definition 6. *A memoryless function* $h : [0, \infty) \times \mathbb{R} \to \mathbb{R}$ *is said to belong to the sector*

- $[0, \infty]$ *if* $uh(t, u) \geq 0$;
- $[K_1, \infty]$ *if* $u[h(t, u) - K_1 u] \geq 0$;
- $[0, K_2]$ *with* $K_2 > 0$ *if* $h(t, u)[h(t, u) - K_2] \leq 0$;
- $[K_1, K_2]$ *with* $K_2 > K_1$ *if* $[h(t, u) - K_1 u][h(t, u) - K_2 u] \leq 0$.

An example of a function that belongs to the sector $[K_1, K_2]$ is given in Figure A.2.

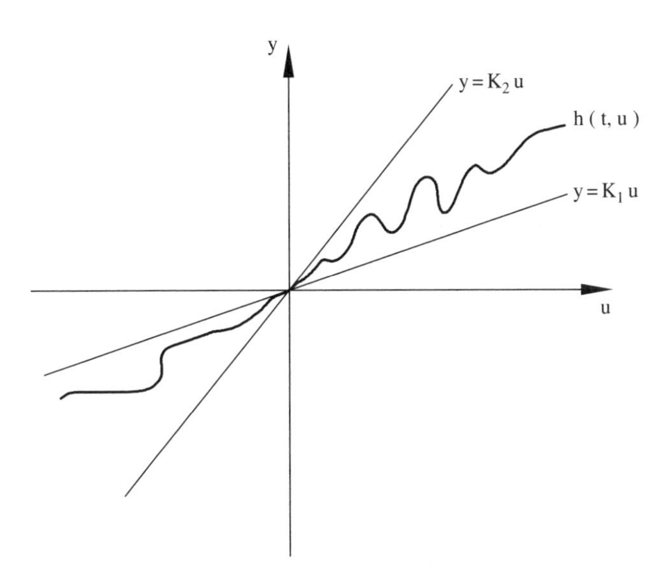

Figure A.2 Example of a function that belongs to the sector $[K_1, K_2]$.

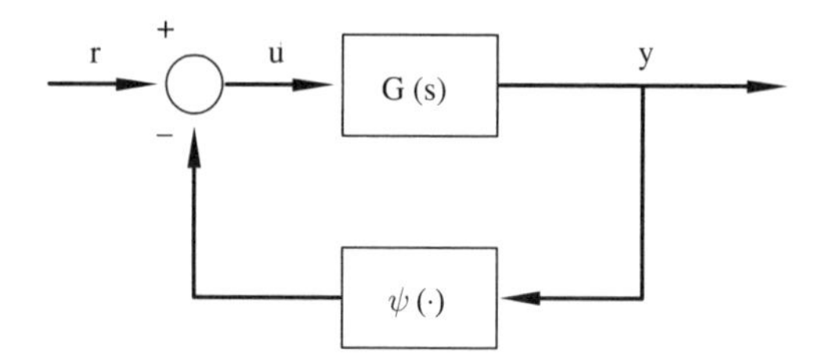

Figure A.3 Equivalent representation of the system (A.19)–(A.21).

A.3.4 Absolute Stability

Consider the feedback connection of Figure A.3. It is assumed that the external input is $r = 0$ and study the behaviour of the unforced system represented by

$$\dot{x} = Ax + Bu \tag{A.19}$$

$$y = Cx + Du \tag{A.20}$$

$$u = -\psi(t, y) \tag{A.21}$$

where $x \in \mathbb{R}^n$, $u, y \in \mathbb{R}$.

It is assumed that (A, B) is controllable, (A, C) is observable and $\psi : [0, \infty) \times \mathbb{R} \to \mathbb{R}$ is a memoryless, possibly time-varying nonlinearity, which is piecewise continuous in t and locally Lipschitz in y. It is also assumed that the feedback connection has a well-defined state model, which is the case when

$$u = -\psi\,(t, Cx + Du) \tag{A.22}$$

has a unique solution u for every (t, x) in the domain of interest. This is always the case when $D = 0$.

The transfer function

$$G(s) = C\,(sI - A)^{-1}\,B + D \tag{A.23}$$

of the linear system is proper. The controllability and observability assumptions ensure that $\{A, B, C, D\}$ is a minimal realization of $G(s)$. From linear system theory, it is known that for any rational proper $G(s)$, a minimal realization always exists. For all nonlinearities $\psi(\cdot)$ satisfying the sector condition, the origin $x = 0$ is an equilibrium point of the system (A.19)–(A.21). The sector condition may be satisfied globally, that is for all $y \in \mathbb{R}$, or satisfied only for $y \in Y$, a subset of $\mathbb{R}$ that is an interval containing the origin.

Definition 7. *Consider the system (A.19)–(A.21), where ψ satisfies a sector condition according to Definition 6. The system is said to be absolutely stable if the origin is globally uniformly asymptotically stable for any nonlinearity in the given sector. It is absolutely stable with a finite domain if the origin is uniformly asymptotically stable.*

Theorem 13. *The system (A.19)–(A.21) is absolutely stable if*

(a) $\psi \in [K_1, \infty]$ and $G(s)/[1 + K_1 G(s)]$ is strictly positive real or
(b) $\psi \in [K_1, K_2]$ with $K_2 > K_1$ and $[1 + K_2 G(s)]/[1 + K_1 G(s)]$ is strictly positive real.

In this case, the origin is globally exponentially stable. If the sector condition is satisfied only on a set $Y \subset \mathbb{R}$, then the foregoing conditions ensure that the system is absolutely stable with a finite domain. In this case, the origin is exponentially stable.

Theorem 14. *Consider the system (A.19)–(A.21), where $\{A, B, C, D\}$ is a minimal realization of $G(s)$ and $\psi \in [\alpha, \beta]$. Then, the system is absolutely stable if one of the following conditions is satisfied, as appropriate:*

(a) If $0 < \alpha < \beta$, the Nyquist plot of $G(j\omega)$ does not enter the disc $D(\alpha, \beta)$ and encircles it m times in the counterclockwise direction, where m is the number of poles of $G(s)$ with positive real parts.
(b) If $0 = \alpha < \beta$, $G(s)$ is Hurwitz and the Nyquist plot of $G(j\omega)$ lies to the right of the vertical line defined by $\mathrm{Re}[s] = -1/\beta$.
(c) If $\alpha < 0 < \beta$, $G(s)$ is Hurwitz and the Nyquist plot of $G(j\omega)$ lies in the interior of the disc $D(\alpha, \beta)$.

If the sector condition is satisfied only on an interval $[a,b]$, then the foregoing conditions ensure that the system is absolutely stable with a finite domain.

A.4 INPUT–OUTPUT PROPERTIES

Lemma 12. *Let $H(s)$ be a stable and strictly proper transfer function, with an input u and an output y; that is $y(s) = H(s)u(s)$. Therefore the following is obtained:*

(*a*) *If $u \in L^1$, then $y \in L^1 \cap L^\infty$, $\dot{y} \in L^1$, y is absolutely continuous and $y(t) \to 0$ as $t \to \infty$.*
(*b*) *If $u \in L^\infty$, then $y \in L^\infty$, $\dot{y} \in L^\infty$ and y is uniformly continuous.*

Lemma 13. *Let f be a differentiable function such that $\dot{f} \in L^1$. Then*

$$\lim_{t \to \infty} f(t)$$

exists and is finite.

Lemma 14. (*Barbalat*). *Consider the function $f : \mathbb{R}_+ \to \mathbb{R}$. If f, $\dot{f} \in L^\infty$ and $f \in L^p$ for some $p \in [1, \infty)$, then*

$$\lim_{t \to \infty} f(t) = 0$$

References

[1] A. Visintin, *Differential Models of Hysteresis*. Springer-Verlag, Berlin, Heidelberg (1994).

[2] P. Duhem, 'Die dauernden aenderungen und die thermodynamik', *I. Z. Phys. Chem.*, **22**, 543–589 (1897).

[3] M. A. Krasnosel'skii and A. V. Pokrvskii, *Systems with Hysteresis*, Nauka, Moscow (1983).

[4] F. Preisach, 'Ber die magnetische nachwirkung', *Zeit. Phys.*, **94**, 277–302 (1935).

[5] J. W. Mack, P. Nistri and P. Zecca, 'Mathematical models for hysteresis, *SIAM Review*, **35**(1), 94–123 (1993).

[6] R. Bouc, 'Modèle mathématique d'hystérésis (a mathematical model for hysteresis)', *Acustica*, **21**, 16–25 (1971).

[7] Y. K. Wen, 'Method for random vibration of hysteretic systems', *Journal of the Engineering Mechanics Division*, **102**(EM2), 246–263 (April 1976).

[8] S. Erlicher and N. Point, 'Thermodynamic admissibility of Bouc--Wen type hysteresis models, *Comptes Rendus Mécanique*, **332**(1), 51–57 (January 2004).

[9] T. T. Baber and Y. K. Wen, 'Random vibration of hysteretic degrading systems', *Journal of Engineering Mechanics*, **107**(6), 1069–1087 (1981).

[10] T. Baber and M. N. Nouri, 'Modeling general hysteresis behavior and random vibration application', *Journal of Vibration, Acoustics, Stress and Reliability in Design*, **108**, 411–420 (1986).

[11] G. C. Foliente, 'Hysteresis modeling of wood joints and structural systems, *Journal of Structural Engineering*, **121**(6), 1013–1022 (June 1995).

[12] J. A. Pires, 'Stochastic seismic response analysis of soft soil sites', *Nuclear Engineering and Design*, **160**, 363–377 (1996).

[13] S. Dobson, M. Noori, Z. Hou, M. Dimentberg and T. Barber, 'Modeling and random vibration analysis of SDOF systems with asymmetric hysteresis', *International Journal of Non-linear Mechanics*, **32**(4), 669–680 (1997).

Systems with Hysteresis: Analysis, Identification and Control using the Bouc–Wen Model
F. Ikhouane and J. Rodellar

[14] C. H Wang and Y. K. Wen, 'Evaluation of pre-Northridge low rise steel buildings. Part I: modeling', *Journal of Engineering Mechanics*, **126**(10), 1160–1168 (2000).

[15] R. Bouc and D. Boussaa, 'Drifting response of hysteretic oscillators to stochastic excitation', *International Journal of Non-linear Mechanics*, **37**, 1397–1406 (2002).

[16] M. H. Shih and W. P. Sung, 'A model for hysteretic behavior of rhombic low yield strength added damping and stiffness', *Computers and Structures*, **83**, 895–908 (2005).

[17] Y. J. Park, Y. K. Wen and A. H. S. Ang, 'Random vibration of hysteretic systems under bi-dimensional ground motion', *Earthquake Engineering and Structural Dynamics*, **14**, 543–557 (1986).

[18] F. Casciati, 'Stochastic dynamics of hysteretic media', *Structural Safety*, **6**, 259–269 (1989).

[19] R. H. Sues, S. T. Mau and Y. K. Wen, 'System identification of degrading hysteretic restoring forces', *Journal of Engineering Mechanics*, **114**(5), 833–846 (May 1988).

[20] C. H. Loh and S. T. Chung, 'A three-stage identification approach for hysteretic systems', *Earthquake Engineering and Structural Dynamics*, **22**, 129–150 (1993).

[21] A. W. Smith, S. F. Masri, A. G. Chassiakos and T. K. Caughey, 'On-line parametric identification of MDOF nonlinear hysteretic systems', *Journal of Engineering Mechanics*, **125**(2), 133–142 (February 1999).

[22] H. Wee, Y. Y. Kim, H. Jung and G. N. Lee, 'Nonlinear rate-dependent stick-slip phenomena: modeling and parameter estimation', *International Journal of Solids and Structures*, **38**(8), 1415–1431 (February 2001).

[23] A. W. Smith, S. F. Masri, E. B. Kosmatopoulos, A. G. Chassiakos and T. K. Caughey, 'Development of adaptive modeling techniques for non-linear hysteretic systems', *International Journal of Non-linear Mechanics*, **37**, 1435–1451 (2002).

[24] S. R. Hong, S. B. Choi, Y. T. Choi and N. M Wereley, 'A hydromechanical model for hysteretic damping force prediction of ER damper: experimental verification', *Journal of Sound and Vibration*, **285**, 1180–1188 (2005).

[25] J. L. Ha, R. F. Fung and C. S. Yang, 'Hysteresis identification and dynamic responses of the impact drive mechanism', *Journal of Sound and Vibration*, **283**, 943–956 (2005).

[26] J. L. Ha, R. F. Fung and C. Han, 'Optimization of an impact drive mechanism based on real-coded genetic algorithms', *Sensors and Actuators A Physical*, **121**(2), 488–493 (30 June 2005).

[27] A. Kyprianou, K. Worden and M. Panet, 'Identification of hysteresis systems using the differential evolution algorithm', *Journal of Sound and Vibration*, **2**(248), 289–314 (2001).

[28] K. H. Hornig, 'Parameter characterization of the Bouc–Wen mechanical hysteresis model for sandwich composite materials by using real coded genetic algorithms', Technical Report, Auburn University, Mechanical Engineering Department, 201 Ross Hall, Auburn, Alabama 36849 (2000).

[29] F. Petrone, M. Lacagnina and M. Scionti, 'Dynamic characterization of elastomers and identification with rheological models', *Journal of Sound and Vibration*, **271**, 339–363 (2004).

[30] I. Pivovarov and O. G. Vinogradov, 'One application of Bouc's model for non-linear hysteresis', *Journal of Sound and Vibration*, **118**(2), 209–216 (1987).
[31] Y. Q. Ni, J. M. Ko and C. W. Wong, 'Identification of non-linear hysteretic isolators from periodic vibration tests', *Journal of Sound and Vibration*, **4**(217), 737–756 (1998).
[32] M. Panet and L. Jezequel, 'Dissipative unimodal structural damping identification', *International Journal of Non-linear Mechanics*, **35**, 795–815 (2000).
[33] S. B. Kim, B. F. Spencer and C. B. Yun, 'Frequency domain identification of multi-input, multi-output systems considering physical relationships between measured variables', *Journal of Engineering Mechanics*, **131**(5), 461–472 (May 2005).
[34] P. Q. Xia, 'An inverse model of MR damper using optimal neural network and system identification', *Journal of Sound and Vibration*, **266**, 1009–1023 (2005).
[35] J. Ching, J. L. Beck and K. A. Porter, 'Bayesian state and parameter estimation of uncertain dynamical systems', *Probabilistic Engineering Mechanics*, **21**, 81–96 (2006).
[36] S. J. Li, Y. Suzuki and M. Noori, 'Identification of hysteretic systems with slip using bootstrap filter', *Mechanical Systems and Signal Processing*, **18**, 781–795 (2004).
[37] S. J. Li, Y. Suzuki and M. Noori, 'Improvement of parameter estimation for non-linear hysteretic systems with slip by a fast Bayesian bootstrap filter', *International Journal of Non-linear Mechanics*, **39**, 1435–1445 (2004).
[38] J. S. Lin and Y. Zhang, 'Nonlinear structural identification using extended Kalman filter', *Computers and Structures*, **52**(4), 757–764 (1994).
[39] S. F. Masri, J. P. Caffrey, T. K. Caughey, A. W. Smyth and A. G. Chassiakos, 'Identification of the state equation in complex non-linear systems', *International Journal of Non-linear Mechanics*, **39**, 1111–1127 (2004).
[40] N. Pastor and A. Rodríguez-Ferran, 'Hysteretic modeling of X-braced shear walls', *Thin-Walled Structures*, **43**, 1567–1588 (2005).
[41] S. J. Li, H.Yu and Y. Suzuki, 'Identification of non-linear hysteretic systems with slip', *Computers and Structures*, **82**, 157–165 (2004).
[42] S. M. Savaresi, S. Bittanti and M. Montiglio, 'Identification of semi-physical and black-box non-linear models: the case of MR-dampers for vehicles control', *Automatica*, **41**, 113–127 (2005).
[43] H. H. Tsang, R. K. L. Su and A. M. Chandler, 'Simplified inverse dynamics models for MR fluid dampers', *Engineering Structures*, **28**(3), 327–341 (February 2006).
[44] N. Gerolymos and G. Gazetas, 'Winkler model for lateral response of rigid caisson foundations in linear soil', *Soil Dynamics and Earthquake Engineering*, **26**(5), 347–361, May 2006.
[45] S. K. Kunnath, J. B. Mander and L. Fang, 'Parameter identification for degrading and pinched hysteretic structural concrete systems', *Engineering Structures*, **19**(3), 224–232 (1997).
[46] F. Ikhouane, O. Gomis-Bellmunt and P. Castell-Vilanova, 'A limit cycle approach for the parametric identification of hysteretic systems', *IEEE Control Systems Technology* (2007) (submitted).
[47] F. Ikhouane and J. Rodellar, 'On the hysteretic Bouc–Wen model. Part II: robust parametric identification', *Nonlinear Dynamics*, **42**, 79–95 (2005).

[48] O. Gomis-Bellmunt, F. Ikhouane, P. Castell-Vilanova and J. Bergas-Jané. 'Modeling and validation of a piezoelectric actuator', *Electrical Engineering* (2007) (to appear).

[49] M. Battaini, K. Breitung, F. Casciati and L. Faravelli, 'Active control and reliability of a structure under wind excitation', *Journal of Wind Engineering and Industrial Aerodynamics*, **74–76**, 1047–1055 (1998).

[50] A. H. Barbat, J. Rodellar, E. P. Ryan and N. Molinares. 'Active control of nonlinear base-isolated buildings', *Journal of Engineering Mechanics ASCE*, **121**(6), 676–684 (1995).

[51] B. M. Chen, T. H. Lee, C. C. Hang, Y. Guo and S. Weerasooriya, 'An H_∞ almost disturbance decoupling robust controller design for a piezoelectric bimorph actuator with hysteresis', *IEEE Transactions on Control Systems Technology*, 7(2), 160–174 (March 1999).

[52] H. Irschik, K. Schacher and A. Kugi, 'Control of earthquake excited nonlinear structures using Lyapunov's theory', *Computers and Structures*, **67**, 83–90 (1998).

[53] F. Ikhouane, V. Mañosa and J. Rodellar, 'Adaptive control of a hysteretic structural system', *Automatica*, **41**, 225–231 (2005).

[54] N. Luo, J. Rodellar and M. de la Sen, 'Composite robust active control of seismically excited structures with actuator dynamics', *Earthquake Engineering and Structural Dynamics*, **27**, 301–31 (1998).

[55] N. Luo, J. Rodellar, M. de la Sen and J. Vehí. Output feedback sliding mode control of base isolated structures', *Journal of the Franklin Institute*, **337**(5), 555–577 (2000).

[56] V. Mañosa, F. Ikhouane and J. Rodellar, 'Control of uncertain non-linear systems via adaptive backstepping', *Journal of Sound and Vibration*, **280**, 657–680 (2005).

[57] F. Pozo, F. Ikhouane, G. Pujol and J. Rodellar, 'Adaptive backstepping control of hysteretic base-isolated structures', *Journal of Vibration and Control*, **12**(4), 373–394 (2006).

[58] F. Ikhouane and J. Rodellar, 'A linear controller for hysteretic systems', *IEEE Transactions on Automatic Control*, **51**(2), 340–344 (2006).

[59] M. D. Symans and M. C. Constantinou, 'Semi-active control systems for seismic protection of structures: a state-of-the-art review', *Engineering Structures*, **21**(6), 469–487 (June 1999).

[60] L. M. Jansen and S. J. Dyke, 'Semi-active control strategies for MR dampers: a comparative study', *Journal of Engineering Mechanics*, **126**(8), 795–803 (August 2000).

[61] A. N. Vavreck, 'Control of a dynamic vibration absorber with magnetorheological damping', Technical Report, Pennsylvania State University, Altoona Collegue, Altoona, Pennsylvania (2000).

[62] G. Yang, B. F. Spencer, J. D. Carlson and M. K. Sain, 'Large-scale MR fluid dampers: modeling and dynamic performance considerations', *Engineering Structures*, **24**, 309–323 (2002).

[63] G. Z. Yao, F. F. Yap, G. Chen, W. H. Li and S. H. Yeo, 'MR damper and its application for semi-active control of vehicle suspension system', *Mechatronics*, **12**, 963–973 (2002).

[64] J. N. Yang and A. K. Agrawal, 'Semi-active hybrid control systems for nonlinear buildings against near-field earthquakes', *Engineering Structures*, **24**(3), 271–280 (March 2002).
[65] Z. G. Ying, W. Q. Zhu and T. T. Soong, 'A stochastic optimal semi-active control strategy for ER/MR dampers', *Journal of Sound and Vibration*, **259**(1), 45–62 (2003).
[66] U. Aldemir, 'Optimal control of structures with semiactive-tuned mass dampers', *Journal of Sound and Vibration*, **266**(4), 847–874 (25 September 2003).
[67] W. Q. Zhu, M. Luo and L. Dong, 'Semi-active control of wind excited building structures using MR/ER dampers', *Probabilistic Engineering Mechanics*, **19**(3), 279–285 (July 2004).
[68] J. G. Chase, L. R. Barroso and S. Hunt, 'The impact of total acceleration control for semi-active earthquake hazard mitigation', *Engineering Structures*, **26**, 201–209 (2004).
[69] R. R. Gerges and B. J. Vickery, 'Design of tuned mass dampers incorporating wire rope springs, part i: dynamic representation of wire rope springs', *Engineering Structures*, **27**, 653–661 (2005).
[70] S. Dong, K. Q Lu, J. Q. Sun, and K. Rudolph, 'Rehabilitation device with variable resistance and intelligent control', *Medical Engineering and Physics*, **27**(3), 249–255 (April 2005).
[71] H. Du, K. Y. Sze and J. Lam, 'Semi-active *H* infinity control of vehicle suspension with magneto-rheological dampers', *Journal of Sound and Vibration*, **283**(3–5), 981–996 (20 May 2005).
[72] N. Luo, J. Rodellar, J. Vehí and M. de la Sen, 'Composite semiactive control of seismically excited structures', *Journal of the Franklin Institute*, **338** (2–3), 225–240 (2001).
[73] D. Badoni and N. Makris, 'Nonlinear response of single piles under lateral inertial and seismic loads', *Soil Dynamics and Earthquake Engineering*, **15**, 29–43 (1996).
[74] B. R. Ellingwood, 'Earthquake risk assessment of building structures', *Reliability Engineering and System Safety*, **74**, 251–262 (2001).
[75] D. Bernardini and F. Vestroni, 'Non-isothermal oscillations of pseudoelastic devices', *International Journal of Non-linear Mechanics*, **38**(9), 1297–1313 (November 2003).
[76] W. Lacarbonara, D. Bernardini and F. Vestroni, 'Nonlinear thermomechanical oscillations of shape-memory devices', *International Journal of Solids and Structures*, **41**(5–6), 1209–1234 (March 2004).
[77] G. P. Warn and A. S. Whittaker, 'Performance estimates in seismically isolated bridge structures', *Engineering Structures*, **26**(9), 1261–1278 (July 2004).
[78] O. Corbi, 'Shape memory alloys and their application in structural oscillations attenuation', *Simulation Modeling Practice and Theory*, **11**(5–6), 387–402 (15 August 2003).
[79] J. Guggenberger and H. Grundmann, 'Stochastic response of large FEM models with hysteretic behavior in beam elements', *Computer Methods in Applied Mechanics and Engineering*, **194**(12–16), 1739–1756 (8 April 2005).

[80] A. X. Guo, Y. L. Xu and B. Wu, 'Seismic reliability analysis of hysteretic structure with viscoelastic dampers', *Engineering Structures*, **24**(3), 373–383 (March 2002).

[81] G. F. Demetriades, M. C. Constantinou and A. M. Reinhorn, 'Study of wire rope systems for seismic protection of equipment in buildings', *Engineering Structures*, **15**(5), 321–334 (1993).

[82] C. A. Schenk, H. J. Pradlwarter and G. I. Schuller, 'Non-stationary response of large, non-linear finite element systems under stochastic loading', *Computers and Structures*, **83**(14), 1086–1102 (May 2005).

[83] P. Saad, A. Al Majid, F. Thouverez and R. Dufour, 'Equivalent rheological and restoring force models for predicting the harmonic response of elastomer specimens', *Journal of Sound and Vibration*, **290**, 619–639 (2006).

[84] S. J. Dyke, F. Yi, S. Frech and J. D. Carlson, 'Application of magnetorheological dampers to seismically excited structures', in *17th International Modal Analysis Conference*, Kissimmee, Florida (1999).

[85] Y. Shen, M. F. Golnaraghi and G. R. Heppler, 'Analytical and experimental study of the response of a suspension system with a magnetorheological damper', *Journal of Intelligent Material Systems and Structures*, **16**, 135–147 (2005).

[86] M. Sasani and E. P. Popov, 'Seismic energy dissipators for RC panels: analytical studies', *Journal of Engineering Mechanics*, **127**(8), 835–843 (2001).

[87] T. S. Low and W. Guo, 'Modeling of a three-layer piezoelectric bimorph beam with hysteresis', *Journal of Microelectromechanical Systems*, **4**(4), 230–237 (1995).

[88] G. Yang, B. F. Spencer, H. J. Jung and J. D. Carlson, 'Dynamic modeling of large-scale magnetorheological damper systems for civil engineering applications', *Journal of Engineering Mechanics*, **130**(9), 1107–1114 (September 2004).

[89] S. B. Choi, S. K. Lee and Y. P. Park, 'A hysteresis model for the field-dependent damping force of a magnetorheological damper', *Journal of Sound and Vibration*, **245**, 375–383 (2001).

[90] F. Yi, S. J. Dyke, J. M. Caicedo and J. D. Carlson, 'Experimental verification of multi-input seismic control strategies for smart dampers', *Journal of Engineering Mechanics*, **127**, 1152–1164 (2001).

[91] M. Constantinou, A. Mokha and A. Reinhorn, 'Teflon bearings in base isolation, ii: modeling', *Journal of Structural Engineering*, **116**(2), 455–474 (February 1990).

[92] B. F. Spencer, S. J. Dyke, M. K. Sain and J. D. Carson, 'Phenomenological model of a magnetorheological damper', *ASCE Journal of Engineering Mechanics*, **123**(3), 230–238 (March 1997).

[93] L. Zaiming and H. Katukura, 'Markovian hysteretic characteristics of structures', *Journal of Engineering Mechanics*, **116**(8), 1798–1811 (August 1990).

[94] M. C. Constantinou and I. G. Tadjbaksh, 'Hysteretic dampers in base isolation: random approach', *Journal of Structural Engineering*, **111**(4), 705–721 (April 1984).

[95] F. Ikhouane and S. J. Dyke, 'Modeling and identification of a shear mode magnetorheological damper', *Smart Materials and Structures*, **16**, 605–616 (2007).

[96] J. Guggenberger and H. Grundmann, 'Monte Carlo simulation of the hysteretic response of frame structures using plastification adapted shape functions', *Probabilistic Engineering Mechanics*, **19**, 81–91 (2004).
[97] J. E. Hurtado and A. H. Barbat, 'Fourier-based maximum entropy method in stochastic dynamics', *Structural Safety*, **20**, 221–235 (1998).
[98] M. Ryu and S. Hong, 'End-to-end design of distributed real-time systems', *Control Engineering Practice*, **6**, 93–102 (1998).
[99] D. Y Chiang and J. L. Beck, 'A transformation method for implementing classical multi-yield surface theory using the hardening rule of mroz', *International Journal of Solids Structures*, **33**(28), 4239–4261 (1996).
[100] C. H. Loh, C. R. Cheng and Y. K. Wen, 'Probabilistic evaluation of liquefaction potential under earthquake loading', *Soil Dynamics and Earthquake Engineering*, **14**, 269–278 (1995).
[101] F. Colangelo, R. Giannini and P. E. Pinto, 'Seismic reliability analysis of reinforced concrete structures with stochastic properties', *Structural Safety*, **18**(2–3), 151–168 (1996).
[102] S. Caddemi, 'Parametric nature of the non-linear hardening plastic constitutive equations and their integration', *European Journal of Mechanics A/Solids*, **17**(3), 479–498 (1998).
[103] J. E. Hurtado and A. H. Barbat, 'Equivalent linearization of the Bouc–Wen hysteretic model', *Engineering Structures*, **22**, 1121–1132 (2000).
[104] J. E. Hurtado and A. H. Barbat, 'Improved stochastic linearization method using mixed distributions', *Structural Safety*, **18**(1), 49–62 (1996).
[105] Y. Q. Ni, Z. G. Ying, J. M. Ko and W. Q. Zhu, 'Random response of integrable Duhem hysteretic systems under non-white excitation', *International Journal of Non-linear Mechanics*, **37**(8), 1407–1419 (December 2002).
[106] N. Mostaghel and R. A. Byrd, 'Inversion of Ramberg–Osgood equation and description of hysteresis loops', *International Journal of Non-linear Mechanics*, **37**, 1319–1335 (2002).
[107] J. Awrejcewicz and L. P. Dzyubak, 'Influence of hysteretic dissipation on chaotic responses', *Journal of Sound and Vibration*, **284**, 513–519 (2005).
[108] M. Noori, M. Dimentberg, Z. Hou, R. Christodoulidou and A. Alezandrou, 'First-passage study and stationary response analysis of a BWB hysteresis model using quasi-conservative stochastic averaging method', *Probabilistic Engineering Mechanics*, **10**, 161–170 (1995).
[109] Z. G. Ying, W. Q. Zhu, Y. Q. Ni and J. M. Ko, 'Stochastic averaging of Duhem hysteretic systems', *Journal of Sound and Vibration*, **254**(1), 91–104 (27 June 2002).
[110] T. Haukaasa and A. D. Kiureghian, 'Strategies for finding the design point in non-linear finite element reliability analysis', *Probabilistic Engineering Mechanics*, **21**(21), 133–147 (April 2006).
[111] K. Breitung, F. Casciati and L. Faravelli, 'Reliability based stability analysis for actively controlled structures', *Engineering Structures*, **20**(3), 211–215 (1998).
[112] P. K. Koliopulos and A. M. Chandler, 'Stochastic linearization of inelastic seismic torsional response: formulation and case studies', *Engineering Structures*, **17**(7), 494–504 (1995).

[113] F. Carli, 'Nonlinear response of hysteretic oscillator under evolutionary excitation', *Advances in Engineering Software*, **30**(9–11), 621–630 (September 1999).
[114] Y. K. Wen, 'Equivalent linearization for hysteretic systems under random excitation', *Transactions of the ASME*, **47**, 150–154 (1980).
[115] N. Okuizumi and K. Kimura, 'Multiple time scale analysis of hysteretic systems subjected to harmonic excitation', *Journal of Sound and Vibration*, **272**, 675–701 (2004).
[116] C. W. Wong, Y. Q. Ni and S. L. Lau, 'Steady-state oscillation of a hysteretic differential model. Part I: response analysis', *Journal of Engineering Mechanics*, **120**(11), 2271–2298 (1994).
[117] C. W. Wong, Y. Q. Ni and J. M. Ko, 'Steady-state oscillations of hysteretic differential model. Part II: performance analysis', *Journal of Engineering Mechanics*, **120**(11), 2299–2325 (1994).
[118] M. Battaini and F. Casciati, 'Chaotic behavior of hysteretic oscillators', *Journal of Structural Control*, **3**(1–2), 7–19 (June 1996).
[119] F. Ma, H. Zhang, A. Bockstedte, G. C. Foliente and P. Paevere, 'Parameter analysis of the differential model of hysteresis', *Transactions of the ASME*, **71**, 342–349 (2004).
[120] C. Y. Yang, H. -D. Cheng and R. V. Roy, 'Chaotic and stochastic dynamics for a nonlinear structural system with hysteresis and degradation', *Probabilistic Engineering Mechanics*, **6**(3–4), 193–203 (1991).
[121] F. Ikhouane and J. Rodellar, 'On the hysteretic Bouc–Wen model. Part I: forced limit cycle characterization', *Nonlinear Dynamics*, **42**, 63–78 (2005).
[122] F. Ikhouane, J. Rodellar and J. E. Hurtado, 'Analytical characterization of hysteresis loops described by the Bouc–Wen model', *Mechanics of Avanced Materials and Structures*, **13**, 463–472 (2006).
[123] F. Ikhouane, V. Mañosa and J. Rodellar, 'Dynamic properties of the hysteretic Bouc–Wen model', *Systems and Control Letters*, **56**, 197–205 (2007).
[124] S. Wiggins, *Introduction to Applied Nonlinear Dynamical Systems and Chaos*, Springer-Verlag, Berlin (1990).
[125] A. F. Filippov, *Differential Equations with Discontinuous Righthand Sides*, Kluwer Academic Publishers (1988).
[126] I. Mayergoyz, *Mathematical Models of Hysteresis and Their Applications*, Elsevier Series in Electromagnetism, Elsevier, Oxford (2003).
[127] W. A. Sutherland, *Introduction to Metric and Topological Spaces*, Oxford Science Publications (1975).
[128] T. T. Baber and Y. K. Wen, 'Stochastic equivalent linearization for hysteretic degrading multistory structures', Civil Engineering Studies SRS 471, Department of Civil Engineering, University of Illinois, Urbana, Illinois (1980).
[129] J. Pires, Y. K. Wen and A. H. S. Ang, 'Stochastic analysis of liquefaction under earthquake loading', SRS 504, Department of Civil Engineering, University of Illinois, Urbana, Illinois (1983).
[130] R. H. Sues, Y. K. Wen and A. H. S. Ang, 'Stochastic seismic performance evaluation of buildings', SRS 506, Department of Civil Engineering, University of Illinois, Urbana, Illinois (1983).
[131] W. Rudin, *Real and Complex Analysis*, 3rd edition, McGraw-Hill, New York (1987).

[132] N. Higham, *Accuracy and Stability of Numerical Methods*, SIAM, Philadelphia, Pennsylvania (2002).
[133] B. F. Spencer, *Reliability of Randomly Excited Hysteretic Structures*, Springer-Verlag, Berlin, 1986.
[134] C. A. Coulomb, 'Théorie des machines simples', *Mémoires de Mathématique et de Physique de l'Académie de Sciences*, **1**, 161–331 (1785).
[135] B. Armstrong-Hélouvry, P. Dupont and C. Canudas de Wit, 'A survey of models, analysis tools and compensation methods for the control of machines with friction', *Automatica*, **30**(7), 1083–1138 (1994).
[136] H. P. Gavin, R. D. Hanson and F. E. Filisko, 'Electrorheological dampers, part I: analysis and design', *Journal of Applied Mechanics, ASME*, **63**, 669–675 (1996).
[137] J. C. Ramallo, H. Yoshioka and B. F. Spencer, 'Smart base isolation strategies employing magnetorheological dampers, *Journal of Engineering Mechanics*, **128**, 540–551 (2002).
[138] P. R. Dahl, 'A solid friction model', The Aerospace Corporation, El-Segundo, TOR-158(3107-18), California (1968).
[139] T. Aguilar, Y. Orlov and L. Acho, 'Nonlinear control of nonsmooth time-varying systems with application to friction mechanical manipulators', *Automatica*, **39**(9), 1531–1542 (2003).
[140] H. Khalil, 'Universal integral controllers for minimum phase nonlinear systems', *IEEE Transactions on Automatic Control*, **45**(3), 490–494 (2000).
[141] C. Byrnes and A. Isidori, 'Limit sets, zero dynamics, and internal models in the problem of nonlinear output regulation', *IEEE Transactions on Automatic Control*, **48**(10), 1712–1723 (2003).
[142] Z. P. Jiang and I. Mareels, 'Robust nonlinear integral control', *IEEE Transactions on Automatic Control*, **46**(8), 1336–1342 (2001).
[143] H. Khalil, *Nonlinear Systems*, Prentice-Hall, Englewood Cliffs, new Jersey (2000).
[144] M. Krstic, I. Kanellakopoulos and P. Kokotovic, *Nonlinear and Adaptive Control Design*, John Wiley & Sons, Inc., New York (1995).
[145] C. A. Desoer and M. Vidyasagar, *Feedback Systems: Input–Output Properties*, Academic Press, New York (1975).

Index

Systems with Hysteresis: Analysis, Identification and Control using the Bouc–Wen Model
F. Ikhouane and J. Rodellar